Math Tools
for
Journalists

Second Edition

Professional/Professor Version

Math Tools
for
Journalists

Second Edition

Professional/Professor Version

By Kathleen Woodruff Wickham

Library of Congress Cataloging-in-Publication Data

Wickham, Kathleen.
  Math tools for journalists / by Kathleen Woodruff
Wickham.-- 2nd ed.
      p. cm.
"Professional/Professor Version."
  ISBN 0-9729937-4-6
  1. Mathematics. 2. Business mathematics. I. Title.
  QA39.3.W53 2003
  510'.24097--dc21

                                    2003009962

Cover design by Michelle Crisanti
Edited by Ed Avis

ISBN 0-9729937-4-6
Printed in U.S.A.

The ISBN for the **Student Edition** of Math Tools
for Journalists - Second Edition is 0-9729937-5-4.
For ordering information, call Marion Street Press,
503-888-4624

Marion Street Press
4207 S.E. Woodstock Blvd. # 168
Portland, Ore. 97206-6267
(503) 888-4624
www.marionstreetpress.com

This book is dedicated to my sons,

Matthew and Timothy,

who make everything worthwhile

Contents

Chapter 7: Stocks and Bonds 101

Chapter 8: Property Taxes 111

Chapter 9: Directional Measurements......121

Chapter 10: Area Measurements 133

Chapter 11: Volume Measurements 141

Chapter 12: The Metric System................147

Foreword

There is no escape. While many reporters and editors sought journalism as a refuge from math, the reality is that numbers are a fact of life in today's newsroom. In a study of 500 news stories, I found that nearly half of the articles involved mathematical calculation or numerical comparison. As the late Victor Cohn of the Washington Post observed, "We journalists like to think we deal mainly in facts and ideas, but much of what we report is based on numbers. Politics comes down to votes. Budgets and dollars dominate government. The economy, business, employment, sports — all demand numbers . . . Like it or not, we must wade in."

There is no escape from math, but the pages ahead will help wordsmiths keep their heads above water. Math Tools for Journalists provides a no-nonsense guide to just about all the mathematics you will encounter covering the news. Kathleen Woodruff Wickham takes the reader step-by-step through calculating percentage change and determining the mean, median and mode when expressing averages. She also delves into statistics, explaining how standard deviation and other higher math applies to journalism. Each topic includes math drills. In short, this book is a handy reference, whether you are a neophyte or a whiz at numbers.

Journalists often shun numbers because they fear them. In focus groups on the use of math in newsrooms, reporters and editors quipped only half-jokingly about the need for "group therapy" to get over their fear of math. Research shows low confidence in math performance even among journalists who had excelled in high school and college-level math. Clearly, many journalists have the aptitude to perform the basic math encountered in day-to-day reporting, yet lack the confidence to put their knowledge to use. Math Tools for Journalists may not be a substitute for group therapy. But by showing step-by-step how to apply math fundamentals to practical journalism, the book helps take the fear out of reporting numbers.

Journalists need help beyond plugging numbers into a mathematical formula. For many, the challenge is knowing how to apply the math in ways that make sense to readers. Demonstrating by example, Math Tools for Journalists provides capsule accounts of how reporters have used mathematics to enrich their stories. In addition, each chapter provides "Learning Challenges" — hands-on activities that take the math lessons beyond textbook applications and into the real world. In doing so, Math Tools for Journalists shows that numbers really are the mortar of good journalism.

Scott Maier
Associate Professor
School of Journalism and Communication
University of Oregon

Introduction

All journalists need basic math skills. Government budgets, crime reports and research statistics form the backbone of daily journalism. Figuring out percentage change, calculating basic time/distance equations and understanding real estate tax assessments are all part of a journalist's daily job.

But journalists as a whole have weak math skills. The evidence is seen every day in the newspaper and on television. Percentages are wrong, there is little understanding of business issues, and polls are reported with misleading statistics.

Ideally, journalists should know basic math before they begin their writing courses. But many do not. The high school years are spent in a flurry of advanced math courses — algebra, geometry, calculus and trigonometry. Required college math courses tend to focus on algebra and logic. Basic math is shunted aside, not used and forgotten from disuse. It's not that journalists can't do basic math. It's that journalists have forgotten how.

Journalists are generally aware of their math deficiencies and frequently develop elaborate ways of writing around the problem. The hide-and-seek method works all too often because their editors also have weak math skills. In other words, the editorial gatekeepers, those charged with ensuring accuracy and completeness in daily news reports, perpetuate the system because of their own weaknesses.

The purpose of this book is to solve this problem in the classroom and in the newsroom.

Math Tools for Journalists is designed to improve the math skills of journalists by providing them with formulas written in language they can understand and with drill problems developed with an eye to their on-the-job experiences.

Each chapter includes a section on problem-based learning, identified as Learning Challenge. Problem-based learning takes the lessons in the chapter outside of the classroom and into the real world with hands-on activities. Students are better able to relate, comprehend, learn and use the material when the information is connected to their area of expertise and assignments are related to the learning task.

Newsroom trainers will also find the book useful. By making the content meaningful to reporters and editors in the newsroom, there is a greater likelihood that the material will be used, retained and referenced as the need develops. To that end, the chapters are written so that the book can also be used as a reference, with the solutions to math questions easy to locate and transferable to on-the-spot problems.

The goal of Math Tools for Journalists is to enable journalists to feel comfortable with math. The end result should be stories written with greater clarity, credibility and accuracy.

The book is also designed for journalism programs seeking to meet the mandates of the Accrediting Council for Education in Journalism and Mass Communication. ACEJMC adopted new standards in Fall 2000 calling on accredited journalism schools to include numeracy skills within the professional values and competencies taught to undergraduate journalism majors. The new principles were effective September 2001. Journalism schools seeking accreditation beginning in 2002-2003 have to utilize the standards as part of the accreditation process.

Acknowledgments

Math Tools for Journalists would not have been published without the interest of Ed Avis of Marion Street Press. Ed took my casual remark about working on a math book seriously. He e-mailed, called me and encouraged me to develop a proposal. Thank you for the encouragement. Without your interest, this book would still be a folder on my computer.

Other people played an important role in Math Tools for Journalists and deserve recognition and thanks.

Appreciation is expressed to Dr. Stuart Bullion, chair of the journalism department at the University of Mississippi, and Dr. Jim Redmond, chair of the journalism department at The University of Memphis.

Stuart is my current boss and deserves recognition because he listened as I struggled with various chapters, provided suggestions and asked good questions about the material, challenging me in the process. He was also very supportive as I battled a severe case of elbow tendonitis from too many hours spent on the keyboard. Stuart devised the course, Writing for the Media, that includes a strong math component. His commitment to improving the basic math skills of journalists served as the impetus to move the book from a conversation to a text. His encouragement and support is deeply appreciated.

Jim is my former boss, my mentor and a friend. In many conversations over a couple of years we worked out the basic structure and philosophy of the book. Thank you Jim for your ideas and the encouragement to go ahead with the book on my own.

Students enrolled in Writing for the Media at Ole Miss from 1999 through 2001 deserve recognition too. Yes, all 300 or so of you. The students enrolled in Writing for the Media were exposed to variations of the math questions in Math Tools for Journalists, were subjected to various teaching approaches and provided feedback regarding the problems, the approach of the text, math and their futures as journalists. I took their comments to heart. The book is better because of their honest appraisal, insights and sincere criticism. Their math and critical analysis skills should be better because of it too.

Appreciation is also expressed to Ole Miss Provost Carolyn Statton for providing funding so that I could attend the Problem-based Learning Conference at Samford University. The skills and techniques of PBL complement my own interest and belief in active learning as an integral part of the learning process.

Thank you to Doug Fisher, a journalism instructor at the University of South Carolina, for his thorough evaluation of the second edition. Many of the improvements of this edition came from his suggestions.

Thank you also to Christopher Karadjov, assistant professor of journalism at SUNY-Oswego, for his wise counsel on the second edition, particularly in the Polls and Surveys chapter.

A broad thank you to all the other journalism professors who suggested improvements and pointed out errors in the first edition. Your comments have made the second edition stronger.

Another important critic of Math Tools was Pat Stith, a reporter at The News & Observer in Raleigh, N.C. Pat's careful critique of the first edition led to many of the changes in the second edition.

For his proofing of the math in this book, thanks goes to mathematician Glen Avis. And for his careful reading of the stocks and bonds chapter, my appreciation is extended to Michael Sirvinskas of William Blair & Company, Chicago.

Three additional people also deserve recognition: Susan Tanner of Bartlett, Tenn., and Ruth and David Cox of Memphis, Tenn. I thank them for their help, their friendship and their skills.

Susan, a math teacher, worked as my editorial assistant during Summer and Fall 2001. Susan copy-edited, reviewed the questions, found the flaws, worked the answers and provided input based on her classroom experiences. Any errors are mine.

Ruth and David, friends and neighbors, took care of my children, fed the dog and offered up tea and conversation whenever I needed an ear and an arm to lean on. Writing a text is a solitary exercise dependent on the good will of others. Without the good will of Ruth and David I would probably still be stuck on Chapter One.

Appreciation is also expressed to my colleagues at the University of Mississippi. From our building overlooking the Grove to coffee at the Union, your support and friendship sustained and inspired me.

Kathleen Woodruff Wickham, Ed.D.
Assistant professor/journalism
The University of Mississippi, Oxford, Miss.

Chapter 1

The Language of Numbers

Numbers are precise. That precision often frightens men and women more comfortable with the verbal options available in the English language. But journalists committed to accuracy in reporting need to understand the language of numbers — the different forms of numbers, how these forms are used to convey information and appropriate usages for words used in combination with numbers.

Look at the front page of the daily newspaper. There are numbers in many articles — from the cost of the war against terrorism to the state of the economy to the funding of schools. You'll also see numbers in articles about education test scores, municipal budgets

Numbers in the news...

Law Has Little Bite

Chicago has 150.000 rabies-vaccinated dogs, but only 18,891 with mandatory dog licenses.

For the first time in its 130-year history of issuing dog licenses, City Hall was supposed to match its list of licensed dogs with Cook County's list of dogs with rabies shots.

The number of licenses has increased, but only by 6,981.

Chicago Sun-Times, Oct. 31, 2002

and air pollution emissions. Using those numbers correctly is a sign of excellence and professionalism.

Several basic rules apply. Start by checking the math of speakers, official reports and budgets. Look for implausible numbers and figures that may have been tweaked to look better. Don't assume the person who prepared the documentation had good math skills. And don't assume that person is above manipulating the numbers to make his or her case stronger. In other words, interview the numbers with the same care that you interview people.

Consider also the form of the information and the words that accompany the numbers. We don't report information using Roman numerals for obvious reasons — average citizens have a hard time understanding them. Along the same lines consider the impact or meaning behind the words chosen to accompany numbers — words such as few/many/several, different from/differ with and more/most.

Numerical illiteracy is not acceptable in the newsroom, or even in society. It is unattractive, slovenly and an unacceptable excuse for poor writing, weak reporting and faulty editing.

Numerical literacy, on the other hand, is the hallmark of professionalism, a foundation for excellence and a signal to our readers that numeracy is important to them.

1.1 Style tips

■ Spell out single digit numbers (one, five, eight).

■ Use numerals for multiple digit numbers (53, 649), unless the number is 1 million or more and you are rounding it to the nearest million or tenth of a million (or billion, trillion, etc.), in which case it is acceptable to use a numeral in front of the spelled out word ($5 million, 1.3 billion tons of coal).

■ Round off larger numbers unless a specific number is

Numerals

A numeral is a symbol that represents a number. There are two basic forms: Roman and Arabic.

Arabic numerals are those most commonly useed in modern society. They include the digits 1, 2, 3, 4, 5, 6, 7, 8, 9, and 0.

Roman numerals are often used for formal titles such as Queen Elizabeth II or Pope John Paul VI, to designate historical events such as wars (World War II) and on buildings and public monuments indicating the date of construction.

The Roman system is based on addition and subtraction and uses letters to express values. The value of a Roman numeral is found by starting at the left of the sequence of letters and adding the values of the succeeding symbols to the right, unless the value is less than that of the symbol to the right. In that case the smaller value is subtracted from the larger value. Only the numbers 1, 10 and 100 can be subtracted, and then only from numbers not more than two steps larger.

One = I	Four = IV	Five = V	Six = VI	Nine = IX
Ten = X	50 = L	100 = C	500 = D	1,000 = M

Examples:

DXIII = 513
500 + 10 + 1 + 1 + 1 = 513

MCMXXCV = 1985
M + (CM) + (XXC) + (V)
1,000 + (1,000 - 100) + (100 - 20) + 5 = 1,000 + 900 + 80 + 5 = 1985

required. For example, the number dead in an airplane accident should always be specific (132), but in many other cases the number can be rounded-off to make it easier for the reader. For example: The Ely Street Department poured about 3,000 tons of salt on town roads last year.

■ Round off numbers to one decimal point whenever possible. For example: The Conley Board of Alderman passed a $4.5 million budget for the next fiscal year.

■ Spell out fractions less than one: one-quarter, two-thirds.

■ When a number is used to assign a position or order in ranking, use words if the number is between first and ninth. If the number is 10 or above use the numeral with its superscript, such as 10th.

■ If a number begins a sentence, it should be written as a word, unless the number is a date, in which case it remains a numeral. Better yet, rewrite the sentence to eliminate the number at the beginning.

■ If corporations or other organizations use numbers as part of their name, follow the corporate style in writing the number in your story. For example: The Big Ten conference led the nation in sacks.

■ Numerals are always used for addresses, dates, highway designations, percentages, speeds, temperatures, times, and weights. Numerals are also always used with money, though numerals and words can be used in combination when the figure is 1 million or more ($3.2 billion). Ages usually are expressed in figures, but some newspapers spell out ages less than 10, especially for inanimate objects (the five-year-old car).

1.2 Writing tips

■ For the sake of clarity, try to limit the number of numbers in each paragraph to no more than two or three. Try to include only one number in the lead of a story.

■ Do the math for your readers. Don't expect the reader to figure out the percentage increase or change over time.

■ Interpret the results in terms the reader can understand. Consider using analogies, storytelling techniques or graphics to illustrate the numbers.

■ Use the word *minus*, not a dash or a hyphen, to avoid confusion.

■ Write out numbers used in slang expressions: *Thanks a million*.

■ In a series, retain the basic style rules even if the numbers are a mixture of single and multiple digit numbers.

1.3. Language tips

■ Among/between: When the items in question are a group or collective, *among* is the appropriate word to use: The University of Michigan fans hid among the Michigan State students after the game. If there are two specific items, or any number of distinct, individual items, *between* is the preferred word: The plane crashed between the houses.

■ Compared to/compared with: To compare one item to another in a figurative usage use *compared to*: Compared to eating dirt, that sandwich was pretty good. To examine two items for their statistical similarities or differences use *compared with*: The chemist compared the results of the second pH test with the results of the third.

■ Different from/different than: *Different from* is the preferred form.

■ Differ from/differ with: When two items are dissimilar use *differ from*: The officers differ from each other in their shooting skills. When two items are in conflict use *differ with*: We differed with the investigator's conclusion.

■ Farther/further: Use *farther* for physical distance and *further* for degree, time or quantity.

■ Fewer/less than: Use *fewer* for items that can be counted: Fewer students failed this semester; fewer apples were eaten than oranges. Use *less than* for mass or time terms: Less than three gallons of paint were used; the strike lasted less than an hour.

■ Fold: Avoid. If you write "a *five-fold* increase" your readers will think you mean to say *five times* something, but technically you are writing that something has *added* five times its original size to its original size, thus it is six times bigger. See *Percentages* and *Times*.

■ Less than/under: *Under* is used to refer to a physical relationship: The bribe was handed to the informant under the desk. Use *less than* to refer to a smaller quantity or amount: He ate less meat than vegetables.

■ More/most: *More* means greater in quantity. *Most* means the greatest in quantity. It is incorrect to use *most* in place of *almost*.

■ More than/over: *Over* is used when referring to spatial relationships. For figures and amounts use *more than*.

■ People/person: Use *person* to refer to one individual

and use *people* for more than one.

■ Percentages: Use percentages in sentences with *more* or *less* or words ending in *-er*. (The company made *10 percent more* money than last year. The diamond was *25 percent bigger* than the other diamond.) See *Fold* and *Times*, and chapter 2.

■ Temperatures: Describe as *higher* or *lower*, not *warmer* or *cooler*.

■ Times: *Times* is a multiplier, and it is best used in a *as much as*-type phrase (The new car cost *three times as much as* the new motorcycle). Don't use *times* with *more* or words ending in *-er*. If the car cost $15,000 and the motorcycle cost $5,000, it would not be correct to write "The new car cost *three times more* than the new motor-cyle" because that implies the reader is supposed to add three times the cost to the cost of the motorcycle (3 x $5,000 *more than* the motorcycle, thus $5,000 + $15,000).

Don't write *times less*, because a figure can't be reduced by more than one times itself. If you want to show a decrease, use percentages.

See *Percentages* and *Fold*.

Number translation

Good journalists translate numerical language into words the average reader can understand.

When you heard the space shuttle Columbia was traveling at Mach 18 when it exploded, did you know how fast the space shuttle was traveling? Most of us only knew the shuttle was moving fast. But some writers took the information and noted that at Mach 18 the shuttle was traveling fast enough to cross the continental United States in 12 minutes.

Here is how they determined that:

1) Mach one is the speed of sound; Mach 18 is 18 times the speed of sound. The speed of sound is 760 miles per hour at sea level and declines at higher altitudes. Columbia exploded at 38 miles, or 200,640 feet, above Earth. At that elevation, the speed of sound is 706 miles per hour based on the formula: $a = \sqrt{g \times R \times T}$ where the speed of sound (a) is equal to the square root of specific heat (g) times the gas constant (R) times the absolute temperature (T). An online calculator at www.grc.nasa.gov assists in computing this formula.

2) Mach 18 x 706 MPH/Mach = 12,708 MPH, so Columbia was traveling 12,708 miles per hour when it exploded.

3) Columbia was scheduled to travel about 2,500 miles from California to Florida. Convert 12,708 MPH to miles per minute by dividing by 60, and you get 211.8 miles per minute. Divide the distance, 2,500, by the speed, 211.8, and you 11.8, or roughly 12 minutes. At 12,708 miles per hour, the shuttle would have taken 12 minutes to cross the United States.

Chapter 2

Percentages

Reporters frequently are faced with figures that could be more clearly explained if they were converted to percentages. By accurately calculating the percentages, the reporter is helping the reader better grasp the issue.

In this chapter we will discuss four common usages of percentages: percentage increase and percentage decrease; percentage of the whole; and percentage points.

Numbers in the news...

Pols and Schools

Ever wonder where your Senators and Representatives send their kids to school?

Well, the Heritage Foundation asked and got answers. Result: 49 percent of the Senators and 40 percent of House members with school-aged children said they send or have sent at least one child to a private school. Not surprising, that includes 33 percent of those who represent our 10 largest cities, where most public schools are in poor shape.

Parade Magazine, Sept. 10, 2000

2.1 Percentage increase/decrease

Formula

Percentage inc./dec. = (new figure - old figure) ÷ old figure

Convert to a percentage by moving the decimal two places to the right

Example

The salary of the Smithtown fire chief was raised from $46,234 to $53,679. What percentage increase was the chief's raise?

New figure = $53,679
Old figure = $46,234

$53,679 - $46,234 = $7,445
$7,445 ÷ $46,234 = .161 = 16.1%

The chief's salary went up 16.1 percent.

The formula for percentage decrease is the same as the formula for percentage increase. The only difference is that your answer will be a negative figure.

Example

Penton Community Chest decided to reduce its donation to the Funtime Preschool from $3,264 to $244. By what percentage was the donation cut?

$244 - $3,264 = -$3,020
-$3,020 ÷ $3,264 = -0.925 = -92.5%

The donation was cut 92.5 percent.

2.2 Percentage of a whole

A million dollars sounds like a lot of money to most people. But if that $1 million is just 1 percent of a city's annual payroll, the figure doesn't seem as big. By calculating the percentage of the whole, a journalist can put things into perspective.

Formula

Percentage of a whole = subgroup ÷ whole group

Move the decimal point two points to the right.

Example

Franklin College spends $1.2 million on its football team. The entire athletic department budget is $3.5 million. What percentage of the entire budget does the football team consume?

$1.2 million ÷ $3.5 million = 0.343 = 34.3%

The football team's budget is 34.3 percent of the entire athletic department's budget.

2.3 Percentage points

When you're working with numbers that are already percentages, it's important to distinguish between *percent* and *percentage point*. One percent is one one-hundreth of something — one percent of $1 is 1 cent. One percentage point might also be one one-hundredth of something — if XYZ Company owns 100 percent of the market for widgets and its marketshare goes down one percentage point, it has lost one percent of its marketshare. But, one percentage point could also be something other than one percent. If XYZ Company

---

Converting fractions to percentages

You can easily convert a fraction to a decimal by completing the division.

13/15 = 13 ÷ 15 = 0.87

In the figure .87, the 8 is in the tenths spot and the 7 is the hundredths spot. Thus, .87 is 8 tenths and 7 hundredths of 1, or simply 87 hundredths of 1. That's very cumbersome — you'd never write that in an article. But you can convert decimals to percentages.

The word *percent* means "of every hundred." For example, if you had 200 peanuts and ate 20 of them, you would have eaten 10 *of every hundred*. Thus, you ate 10 percent of the peanuts.

In the example above, .87 means 87 hundredths. This can also be expressed as 87 "of every hundred," or simply 87 percent.

To quickly convert a decimal to a percentage, just slide the decimal point two points to the right.

Sports stats

If you read the sports pages, you know percentages are the mainstay of many sports statistics. Baseball statistics tend to be a bit more complicated than most other sports'.

■ Batting Average: Take the total number of at-bats and subtract the walks, the sacrifices and the number of times the batter was hit by a pitch. Now divide the number of hits the batter made (where he was safe on a base) by the whole. This gives you the batting average. Example: Nate Avis got up to bat 120 times during the season. He was walked 13 times, hit two sacrifice flies, and got hit by 5 pitches. He hit safely on 40 at-bats.

120 times to bat - 20 walks/sac./hit by pitch = 100 at-bats
40 hits ÷ 100 at-bats = .400 batting average

■ Slugging Percentage: This figure is similar to batting average, but it assigns more importance to extra-base hits. Take the total number of at-bats minus walks, sacrifices and hit-by-pitches. Now add up all the base hits, but count four points to a home run, three to a triple, and so on. Now divide that number by the at-bats. Using the same example from above, let's say 20 of Avis' 40 hits were singles, 10 were doubles, 2 were triples and 8 were homers.

(4 x 8 homers) + (3 x 2 triples) + (2 x 10 doubles) +
(1 x 20 singles) = 78
78 ÷ 100 at bats = .780 slugging percentage

■ Earned Run Average: Earned Run Average is a measure of how often a team scored against a particular pitcher. The formula is ERA = R(27/ip), where R = the number of runs scored during a pitcher's time on the mound (not counting runners who were already on base when the pitcher came in) and ip = the number of outs made while the pitcher was on the mound. Example: Jose Morales pitched three complete innings, and two runners scored during those innings.

ERA = 2(27/9) = 2(3) = 6.00 earned run average.

holds only 5 percent of the market for widgets and loses one *percentage point*, its new marketshare is 4 percent, which is 20 percent lower than a 5 percent marketshare.

Example

The September unemployment rate was 5.6 percent. The October unemployment rate was 7.4 percent. By how many percentage points did the rate go up?

7.4% - 5.6% = 1.8 percentage points

The rate climbed by 1.8 percentage points. If you want to calculate the **percentage increase**, simply follow the formula for percentage increase, section 2.1, using the above figures.

1.8 ÷ 5.6 = .32 = 32%

The unemployment rate climbed 32 percent between September and October.

What if the rates went down instead of up? The formula remains the same. Let's say the June unemployment rate was 11.3 percent. The July unemployment rate was 9.4 percent.

9.4 - 11.3 = -1.9
-1.9 /11.3 = -16.8%

The unemployment rate dropped 1.9 **percentage points** and 16.8 **percent** between June and July.

2.4 Simple/annual interest

A common use for percentages is in computing interest. This section deals with simple interest, which is the most basic form.

The amount of money borrowed is called the *principal*. Money paid for the use of money is called *interest*.

The rate is the percent charged for the use of money. The amount of interest charged depends on the length of time the borrowed money is kept. Interest rates vary but most are calculated based on one year.

Formula

Interest = principal x rate (as a decimal) x time (in years)

Example

Catherine Jason, a new reporter for the Bingham Bugle, borrowed $1,600 from the Thomasville National Bank to make a down payment on a car. She agreed to pay 5 percent interest, payable in one payment at the end of the year. What was her interest payment?

$1,600 x .05 x 1 year = $80

It's simple to modify the formula to work with shorter time periods. Either work the problem using the 12-month formula and multiply the solution by the fraction of the year represented by the shorter time period, or simply divide the rate by that fraction.

. Using the example above, say Catherine pays back the loan at the end of three months. Three months is one quarter of a year, or .25.

$80 x .25 = $20

Or, multiply the 5 percent rate by .25 to get 1.25 percent. Plug that into the equation:

$1,600 x .0125 = $20

2.5 Compounding interest

"Compounding" means the interest is added to the original principal, and subsequent compoundings

---

apply the interest to the principal plus the interest of the previous compoundings.

For example, if a bank were to compound interest on a loan annually, it would calculate the interest on the principal (the original amount of the loan) once each year and add that to the balance. The next year it would calculate the interest based on the remaining principal and interest and add that to the balance, and so on.

Often loans are compounded more frequently than annually. If a loan is compounded monthly, the bank calculates the interest owed on the loan for that month and adds it to the balance. (It calculates the interest based on 1/12th the annual interest rate, since it's only for that one month.)

A loan that is compounded monthly will cost the borrower more than a loan that is compounded annually, even though the rate is the same. This difference is because each compounding adds a small amount to the balance, which makes the next compounding a bit higher, and on and on.

For example, let's say a person borrows $1,000 from the bank at an interest rate of 12 percent. She plans to pay the entire balance in exactly one year (that is, she's not making any payments during the year). If the bank compounded the interest annually, the day she walked into the bank to pay off her loan it would cost her $1,120, because 12 percent of $1,000 is $120.

Now let's say the interest is compounded monthly instead of annually. At the end of the first month, the bank would multiply the principal, $1,000, by 1 percent and add that to the balance (1 percent is 1/12th of 12 percent). So her balance at the end of that first month would be $1,010, because 1 percent of $1,000 is $10. The next month the bank would multiply the new balance, $1,010, by 1 percent, and add that to the balance. One percent of $1,010 is $10.10, so the new balance would be $1,020.10. The monthly compounding would happen 10 more times in the year. At the end of the year when the borrower came in to pay off her loan, the balance would be $1,126.83. This is $6.83 more than the bank would charge if it compounded interest annually.

Payments on loans

Consumers commonly make monthly payments on their loans. Home mortgages and car loans, for examples, almost always require monthly payments. If you know the original loan amount, the interest rate, and the term of the loan (how long the borrower has to repay it), you can calculate the monthly payment and the total interest paid.

As with other compounded loan calculations, these numbers are more easily figured with an advanced calculator or online calculator. But here's the formula in case you want to calculate it yourself.

Formula

A = monthly payment
P = original loan amount
R = interest rate, expressed as a decimal and divided by 12
N = total number of months

$$A = [P \times (1 + R)^N \times R] \div [(1 + R)^N - 1]$$

(The little N in the air beside the brackets means *to the power of*, which means you multiply the result inside the brackets by itself N number of times. For more on math basics, see Appendix 1, page 161).

Example

Acme Press Co. bought a warehouse for $100,000. It financed the entire amount at 6 percent for a period of 10 years. What will its monthly payment be?

$$A = [\$100,000 \times (1 + .006)^{120} \times .006] \div [(1 + .006)^{120} - 1] = \$1,110.50$$

So Acme's monthly payment would be $1,110.50. Perhaps your readers would like to know the total

---

amount of interest Acme will ultimately pay for this warehouse. Just multiply the monthly payment by the number of payments — in this case 120 — and then subtract the original principal amount.

120 x $1,110.50 = $133,260 - $100,000 = $33,260

Interest on savings

Savings accounts and certificates of deposit generally pay compound interest.

Formula

B = balance after one year
P = principal
R = interest rate
T = number of times per year the interest is compounded

$B = P(1 + [R \div T])^T$

Example

Lizzie Gibbons, a reporter for the Easthampton Eagle, decided to combine her $300 income tax refund with $1,700 in her savings account to buy a $2,000 certificate of deposit. The bank awarded Lizzie an 8 percent interest rate, compounded monthly. How much interest did she earn in one year?

$B = \$2000 \ (1 + [.08 \div 12])^{12} = \$2166$
$2166 - $2000 = $166 interest

Skill Drills

1. Alpha police announced a record 8,294 arrests for driving under the influence (DUIs) last year. The previous year, records indicate there were 6,759 arrests. What is the percentage increase in DUI arrests?

2. The Madison City Council budgeted $5,283 for snow removal this year. Last year, Madison budgeted $14,700 for snow removal. What is the percentage decrease?

3. Lisa Palmyra, the sports editor at the Mt. Moriah Times, wants to figure out what percentage of female athletes from the United States earned at least one medal at the summer Olympics in Australia. She knows 309 female athletes participated in the games and 227 earned at least one medal. What percentage earned medals?

4. New Mexico's annual growth rate dropped from 5 percent to 3 percent. Express the drop in both percentage points and percentage.

5. The Brookhaven City Council budget for last year shows that the police department accounted for 36.1 percent of the overall municipal budget. This year, the police department accounts for 34.3 percent of the budget. By how many percentage points did the police department's portion of the budget drop? What is the percent decrease?

6. The nonprofit Hudson Municipal Arts Center had a budget of $1 million two years ago, and it generated 67 percent of that from admissions and concessions. The balance of its budget was covered by the county government. Last year its budget was again $1 million, but it only earned 54 percent of that from admissions and concessions. By how many percentage points did its revenue from admissions and concessions drop? What percentage drop was that? How much more did the county have to pay last year than two years ago? What percentage increase was that?

7. The mayor of Oakville announces that 4/5 of the village's streets were cleared within a day of a major snowstorm. What percentage of the streets was cleared?

8. Amanda Wolfe took out a school loan for two semesters (10 months). How much interest must Amanda pay on a loan of $6,300 at 6.5 percent interest, compounded annually, assuming she pays back the loan in total at the end of the 10 months?

9. Cindy Willis is a musician for the On the Circle Players, a local theater group. She is currently paid $24,000. She negotiated a new salary that calls for a 2.5 percent increase the first year, a 3.5 percent increase the second year and a 4 percent increase the third year, based on her current salary. How much will Willis be making in the third year of the contract?

10. The City of Macintyre financed a new $75,000 snowplow at 3.4 percent for five years. What are the monthly payments? What will be the total interest the city pays?

Learning Challenge

Consider the following stories. These assignments could be written as news stories with additional interviews, presented as charts and/or developed into time lines.

1. Where do the politicians in your town, county or state send their children to school? Are they attending public, private or parochial schools? What about colleges? Are the elected officials in the state who approve higher education budgets sending their children to universities in state or out of state?
You can limit the story to elected officials with children currently in school, or you can draw comparisons from five or 10 years ago. The answer may be a measure of how the people who control funding for schools feel about education.

2. Track the unemployment rate of your city or county over five years. Calculate the differences between years.

3. Examine trends in college tuition, room and board costs over decades. How has each segment fluctuated over the years?

4. Pick three majors at your college and determine the number of graduates last year in each major. Determine the percentage of the total class of each major, then compare that percentage to percentages from 10 years ago. Have they changed? Why?

Chapter 3

Statistics

Next to working with percentages, the most common numbers reporters regularly encounter involve statistics. They are used to report crime rates, the average cost of food at the grocery store and school student test scores. Statistics also are used in research to make accurate inferences about the topic under study.

Having a basic understanding of statistics and the role played by the manipulation of numbers is an important element in a journalist's toolbox. Journalists are frequently asked to evaluate surveys and studies; unless they know how the numbers were used they cannot report accurately on the results.

Remember that statistics can easily be manipu-

Numbers in the news...

City Schools Rank

Memphis City Schools rank last academically among Tennessee's 137 public school districts....

Lee McGarity, with the Memphis district's office of research and evaluation... complained that Roger Abramson, research and policy analyst, averages the percentile scores to come up with overall rankings, which she said skews the statistics....

The Commercial Appeal, Memphis, Tenn., March 11, 2001

lated to serve someone's goal. In the article excerpt on the previous page, for example, an analyst averaged percentile scores. This may have been done to make the numbers fit the analyst's needs, as Mr. McGarity accused. Beware of such manipulations, and report them so readers can make their own decisions.

3.1 Mean

The mean is the sum of all figures in a group divided by the total number of figures. The mean is commonly called the "average."

Example

The following are salaries for midlevel computer technicians in the Midwest. What is average/mean salary?

A Corp: $45,600
B Corp: $54,000
C Corp: $47,800
D Corp: $62,400
E Corp: $50,500
F Corp: $48,000
G Corp: $64,600

Added together the figures = $309,200
$309,200 ÷ 7 = $53,271.43 is the mean.

3.2 Median

Median is the midpoint in the grouping of numbers. To determine the median, rewrite the numbers from lowest to highest; find the number that has the same number of numbers above and below it. If you have an equal number of figures in the group and the two middle numbers are the same, that is the median. If you have an equal number of numbers and the two middle numbers are not the same, add those two middle numbers together and divide by two. That is the median.

Example
Find the median of the Midwest computer technician salaries.

Rearranged from lowest to highest, the salaries are:

$45,600
$47,800
$48,000
$50,500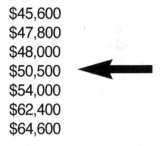
$54,000
$62,400
$64,600

The salary in the middle, the median, is $50,500.

3.3 Mode
The mode refers to the number appearing most frequently in a distribution of numbers. When the numbers are plotted on a graph, high points generated by the collection of numbers (the mode) may indicate an area of interest calling for further investigation or explanation. To determine the mode just count each time each number appears. If each number appears only once there is no mode. If several numbers appear the most they are all modes.

Example
The following are the number of eggs hybrid chickens laid in a given month at the Jonesboro Research Farm. What number is the mode?

12 18 24 25 24 11 28 24 25

The most frequent number to appear, the mode, is 24.

When do you use mode, median or mean?

Mode, median and mean each has a place, and it's up to the researcher or journalist to decide which figure most accurately reports the story.

Let's say you are writing about entry-level salaries of engineers. You interview 12 beginning engineers and learn that 11 of them earn $50,000 per year, and one, a superstar, earns $100,000 per year. In this case it would be appropriate to report the *mode*, $50,000, since that most accurately reflects the salaries of entry-level engineers.

Now let's say three of the engineers earn $45,000, five earn $50,000, three earn $60,000, and one earns $70,000. In this case it would be appropriate to report the *mean* salary, $53,000. Even though none of them earns *exactly* $53,000, that figure accurately portrays entry-level engineer salaries.

Finally, let's say four of the engineers earn $45,000, three earn $50,000, three earn $55,000, one earns $70,000 and one, the son of the boss, earns $350,000. The mode would be $45,000, which doesn't tell your readers much. Because of the large top salary, the mean, $76,250, would be too high to accurately reflect the salaries. So in this case the median, $50,000, would most accurately represent the salaries.

3.4 Percentile

A percentile is a number that represents the percentage of scores that fall at or below the designated score. A test-taker with a score in the 75th percentile knows that 75 percent of those who took the test scored the same or lower than he did. A percentile score is based on its relationship to all the other scores. For example, a person who scores 95 percent on a test would probably be happy with his performance, but his percentile rank would be low if everyone else taking the test scored 96 percent.

Formula

Percentile rank = (Number of people at or below an individual score) ÷ (number of test takers)

This formula can easily be turned around to tell you the number of people who scored at or below a certain point, if you know the percentile rank:

Number of people scored at or below that level = (Percentile) x (number of test takers)

Example

Joe Cadle received the results of his state achievement tests. He received an overall score of 91. Joe found from reading the brochure that came with the results that 1,600 students took the test. Joe's score is equal to or higher than the scores of 1,489 students. What is Joe's percentile rank?

1,489 ÷ 1,600 = 93rd percentile

You know your percentile rank is 65, and 139 people took the test. How many people scored the same or lower than you?

.65 x 139 = 90

Ninety people scored at or below the 65th percentile.

3.5 Standard deviation

Standard deviation is a figure one commonly sees in scientific reports, investment documents and other reports using statistics. Standard deviation indicates how much a group of figures varies from the norm.

A small standard deviation means the figures are consistently grouped around the mean. For instance, a scientific experiment that consistently shows the same results will have a low standard deviation, which may tell the researcher his experiment is valid. Conversely, a high standard deviation may tell him he is getting inconsistent results.

Statistical data are frequently viewed graphically in a bell curve. The middle of the curve — the highest point — is the mean, and the rest of the "bell" spreads out on each side, representing the spread of all the other numbers in the group. The steeper the bell, the smaller the standard variation, since a larger number of numbers are close to the mean. A broad bell represents a large standard variation.

Standard deviation can be used as unit of measure along the bell curve. For example, let's say the standard deviation on a series of stock prices is found to be $2, and the mean of that series of $12. In that case, one positive standard deviation away from the mean would be $14, and one negative standard deviation away from the mean would be $10.

Standard deviation values are written in the same terms as the original data (e.g., points, dollars, etc.)

If the data exhibit a typical distribution, 68 percent of the scores will fall within one standard deviation (either positive or negative), 95 percent will fall within two standard deviations, and 99 percent will fall within three standard deviations.

Standard deviation is a rather complicated figure to compute, and journalists rarely are asked to do so. But it may be helpful to know how.

Formula

Subtract the mean from each score in the distribution.

Square the resulting number for each score.

Compute the mean for these numbers. This figure is called the variance.

Find the square root of the variance.

Example

Public school children in Hanson County can either enroll in the Jordan schools or the Hickory schools. Education reporter Lois Gagan decided to compare the test scores of both school districts in order to write a comparison report.

School	Jordan	Hickory
PS 1	45	30
PS 2	67	77
PS 3	52	32
PS 4	54	74
PS 5	88	91
PS 6	71	86
PS 7	74	74
PS 8	87	72

Scores in the Jordan district range from 45 to 88. Scores in the Hickory district range from 30 to 91. The mean, or average, for both was 67. The standard deviation is needed to determine if there is any significant difference between the two districts.

To calculate, determine each school's deviation from the mean by subtracting the mean from each individual school score.

School	Jordan	Hickory
PS 1	-22	-37
PS 2	0	10

PS 3	-15	-35
PS 4	-13	.7
PS 5	.21	.24
PS 6	.4	.19
PS 7	.7	.7
PS 8	.20	.5

Square the deviations:

School	Jordan	Hickory
PS 1	484	1,369
PS 2	0	100
PS 3	225	1,225
PS 4	169	49
PS 5	441	579
PS 6	16	361
PS 7	49	49
PS 8	400	25

Add the squared deviations and divide by the number of units/scores to get the variance.

	Jordan	Hickory
Sum	1,784	3,754
Divided by 8	223	469.25

Find the square root to get the standard deviation.

	Jordan	Hickory
Standard deviation	15 (rounded)	22 (rounded)

Although the average scores for the two districts are the same, the scores for Jordan are closer to the mean than the scores for Hickory. This is evident by the smaller standard deviation.

What does this mean to your readers? It means there is a greater diversity of performance among students at Hickory School. It would take further investigation to determine why that is, but it could mean the education at Jordan is more consistent than at Hickory, or that Hickory draws its student body from a more intellectually diverse community.

3.6 Probability

Lottery numbers, traffic accidents and fatal illnesses often make the news. In many cases stories like this
could benefit from an understanding of probability.

There are entire textbooks devoted to probabilities,
but calculating a probability simply boils down to a
ratio. For example, according to the American Heart
Association, nearly 2,600 Americans die of cardiovascular disease each day. That's a big number, but let's put
it into perspective. There are about 290 million people
in the United States, so the odds of any one person
dying on a particular day of cardiovascular disease
would be:

2,600 deaths ÷ 290 million people = .0000089

That's an unwieldy number. It's more helpful to
use the phrase "one out of" to describe such a probability. To do this, divide 1 by the answer.

1 ÷ .0000089 = 112,000

So the odds of any given person dying of cardiovascular disease on any given day is "one out of
112,000."

Another common way to show health-related figures like this is to show the number of occurrences "per
100,000 people."

Formula
Deaths per 100,000 people = (Total deaths ÷ total
population) x 100,000

Example

Let's use the case above, but instead of daily figure
let's use an annual figure. About 950,000 Americans die
of cardiovascular disease each year.

Deaths per 100,000 people = (950,000 ÷ 290 million) x 100,000 = .0033 x 100,000 = 330

So, annual U.S. deaths by cardiovascular disease are about 330 per 100,000 people.

There is an important caveat to remember when doing probability stories regarding illnesses: In most cases, they are not completely random events. For example, while the cause of cardiovascular disease has not been pinpointed, there are many known risk factors.

In the example above, we used the entire United States population to calculate the death rate by cardiovascular disease per 100,000 people. But in reality, the odds are not equal for the entire population — very few children die of it, and according to the American Heart Association, people who smoke have twice the risk of heart attack as non-smokers.

Thus, in order to create a more accurate probability article for cardiovascular disease, it would be wise to narrow your population by risk group.

About 25 percent of the American population is under age 18, which means there are about 218 million adult Americans. Assuming that all cardiovascular deaths happened among adults, the rate per 100,000 adult Americans becomes:

(950,000 ÷ 218 million) x 100,000 = 436

This figure is slightly more accurate than the one above, but we can narrow it even further by taking smoking into account.

About 25 percent of the adult population in America smokes, or about 55 million people. These people have twice the risk of non-smokers.

Finding the rate for smokers requires a bit of algebra. Let's say the rate for non-smokers is X. Then the rate for smokers is twice that, 2X. There are 163 million

The language of risk

Articles involving the probability of illness or death often scare readers. To avoid needless worry, ask yourself these questions, which were developed by the Social Issues Research Centre in England.

■ Have the findings been published in a peer-reviewed journal?

■ Do the researchers have an established track record in the field and are they based at a reputable institution or organization?

■ What are the affiliations of the researcher(s)?

■ What do other professionals in the field think of the methods?

■ Are the findings preliminary or inconclusive?

■ Do the findings differ markedly from previous studies?

■ Do these findings appear to contradict mainstream scientific opinion?

■ Are these findings based on small or unrepresentative samples?

■ Do these conclusions generalize to humans from animal studies?

■ Have the researchers only found a statistical correlation?

■ Has the risk been expressed in absolute as well as relative terms?

■ Can the risk be compared with anything else?

■ Will the report cause undue anxiety or optimism among audiences or readers?

■ Have important caveats been prominently included?

■ What do specialist journalists think about the report?

■ Is the headline a fair reflection of the report?

■ What do other professionals in the field think of the research?

adult non-smokers, and 55 million adult smokers. There are 950,000 cardiovascular deaths among both groups. Thus,

$$(163,000,000)X + (55,000,000)2X = 950,000$$

$$163,000,000X + 110,000,000X = 950,000$$

$$274,000,000X = 950,000$$

$$X = 950,000 \div 274,000,000 = .00347$$

$$.00347 \times 100,000 = 347 \text{ deaths per } 100,000$$

So, the new rate for adult non-smokers is 347 deaths per 100,000, and the rate for adult smokers is double that, or 694 deaths per 100,000. This figure is more meaningful than the first we came up with, because it takes into account age and smoking, two critical factors in cardiovascular disease. Now you know why insurance rates are so much higher for smokers!

Lottery = pure chance

Some probability issues are cleaner. Winning the lottery, for example, is pure chance.

Articles profiling lottery winners often give the odds involved in winning the top prize. How do you calculate these odds? Fortunately, in most cases, they are already calculated for you by the authority issuing the lottery. But it's good to know the math behind the odds.

Let's start with something simpler: a coin toss. A coin has two sides, so the odds of getting heads on one toss are one out of two, written as 1/2. If you tossed the coin twice, there would be four possible combinations: head/tail, head/head, tail/head or tail/tail.

Thus, the odds of getting two heads would be 1/4. (It is not a coincidence that the odds of getting two heads in two tosses is double the odds of getting one head in one toss. For any series of events, the odds of a specific outcome equal the odds of each event multiplied together. If you remember fractions [see Appendix 1], you know that $1/2 \times 1/2 = 1/4$.) That basic calculation holds up no matter how many choices there are.

Formula

Odds of a series of events = Odds of first event x odds of second event x odds of third event.....

If the odds are the same for each event in a series of events — say, 10 coin tosses in a row — you can simplify the calculation by using the "to the power of" function. The odds of getting heads on 10 consecutive coin tosses would be $1/2 \times 1/2 \times 1/2$ and so on 10 times. This is the same as 1 divided by 2 to the 10th power, or 1/1,024.

Formula

O = Odds

N = Number of events

Odds of a series when each is the same = O^N

Example

The odds of guessing a random card pulled from a standard deck is 1/52. The odds are the same each time you guess, so what would be the odds of guessing the card three times in a row?

Odds = $1/52^3$ = 1/140,608

Lotteries are more complicated than coin tosses and card games. The Big Game, a lottery sponsored by six states in the eastern United States, requires players to choose five numbers between one and 50, and then choose one "big money ball" number from one to 36 (this ball is drawn from a different drum than the other six balls). Thus, the winner must choose the right number from one to 50 five times, and from one to 36 once.

Let's break the problem down. Let's say the lottery had only one number from one to 50. If you were allowed to guess just one number, then the odds of winning would be one out of 50, written as 1/50. Now let's say the lottery uses two numbers from one to 50, and you choose two unique numbers. Remember, the lottery doesn't repeat numbers, and it doesn't matter what order you choose the numbers in. So, the odds of getting one of the numbers right when the first ball is picked is 2/50 (because you choose two balls, and there are 50 balls in the drum), and the chance of getting the second number right is 1/49 (if you got the first number right, you only have one choice left but there is one fewer ball in the drum). Thus, the odds of correctly choosing the winning number in a two-number lottery would be:

$$2/50 \times 1/49 = 2/2,450 = 1/1,225$$

Your odds are one out of 1,225.

Now let's return to the Big Game. Remember that for the first five numbers, you choose five unique numbers, and the drum starts out with 50 balls. So each time you get a number right, there is one fewer guess for the next ball and one fewer ball in the drum. To correctly guess all five numbers of the main part of the lottery, the odds would be:

$$5/50 \times 4/49 \times 3/48 \times 2/47 \times 1/46 = 120/254,251,200$$

Now, remember that the "big money ball" comes from a different drum, and there are 36 balls in that drum. You only choose one number. So your odds are

1/36. Add that to the calculation above and you get:

120/254,251,200 x 1/36 = 120/9,153,043,200

You would prefer the odds to be "one out of" something, so you need to reduce that fraction to 1 over something. To do that simply divide both the top and bottom by 120. This gives you 1 on the top, and 76,273,360 on the bottom. So, the odds of winning the Big Game are one out of 76,273,360.

Skill Drills

1. The Chandler Police Department keeps a yearly record of the different number of arrests in various categories. Following are the reported arrests for DUI for the past 10 years. Calculate the mean:

2001: 7   2000: 92   1999: 53   1998: 81   1997: 46

1996: 19   1995: 34   1994: 51   1993: 6   1992: 25

2. Workers at the Overstreet Sewing Factory are paid varying rates based on skill levels, seniority and responsibilities. What is the mean pay at the factory for all positions? The number in parentheses represents the number of workers on one shift at the factory.

Detail workers (18 workers)—$7.98 an hour
Senior detail workers (2 workers)—$9.98 an hour
Large item workers (6 workers)—$5.48 an hour
Senior large workers (4 workers)—$7.48 an hour
Level One supervisor (2 workers)—$11.49 an hour
Level Two supervisor (1 worker)—$15.35 an hour

3. Michael Bailey, a technology writer for the Park Project, surveyed companies offering Internet access in Park. What was the median price?

Jaynes Wireless—$4.95 a month
Platt Cellular—$9.95 a month
Holt Online—$19.95 a month
Goodwin Phones—$8.95 a month
Holliday Wireless—$18.95 a month
Phoebe Phones—$28 a month
Wiseman Cellular—$12.28 a month
Branim Wireless—$21.95 a month

Norton Cellular—$16.75 a month
Anderson Online—$7.98 a month

4. Jack Garfield is a patrolman for the Henderson Police Department. He scored in the 60th percentile on the sergeant's exam. He knows that 116 patrolmen took the test. Only the top 30 scorers will be promoted to sergeant. Did he make it?

5. Reid Western conducted a study of online writing classes. He gave 356 students at four colleges who took the courses an exam. Students at Rheingold averaged 56, students at Buchanan averaged 82, students at Bell averaged 76, and students at Online U. averaged 69. The overall average for the four groups was 71. What was the standard deviation?

6. The prospectus for Gold Platter Mutual Fund states that the mean monthly rate of return over the past 36 months was 2 percent, and the standard deviation was .01 percent. The mean return for the Rocksolid Mutual Fund was 2.2 percent during that period, and its standard deviation was .25 percent. Which fund do you think is the most volatile? If you were hoping to get rich quick, which fund would you most likely invest in? If you were worried about risking your investment, which fund would you likely prefer?

7. You read police reports showing that on average, five people per year are murdered in Fitztown. The population of Fitztown is 540,000. What are the odds of any given resident being murdered in a year? You investigate further and learn that no one under age 18 has ever been murdered, and, on average, four out of five murders have occurred in Fitztown's rowdy red light district. Census data show that 25 percent of the population of Fitztown is under age 18, and a survey by your newspaper's entertainment section showed that about one out of 20 residents frequent the red light district. What are the odds of an adult who never visits the red light district getting murdered?

Learning Challenge

Statistics are mathematical tools for interpreting what data mean and for forecasting future trends based on those findings. This section delved into descriptive statistics, which describe data without trying to made predictions or inferences from them. Chapter 5, Polls and Surveys, includes material on inferential statistics, which are based on random sampling and are used to predict or infer things about the entire population under study.

The key to using descriptive statistics in news stories is to make the stories meaningful to the reader and the number an integral part of the story. At the same time the writer must avoid excessive detail that will confuse the reader and bog down the narrative.

1. Obtain the state school test scores for your local district. How do the scores compare to scores in the state's largest school district? How do the scores compare to the scores in the state's smallest school district? Write your results in the form of a news story.

2. Go online and find the standard deviation of three different mutual funds. Describe the relative risk among the three.

3. Learn about your state's lottery (or another state's, if your state doesn't have one). Calculate the odds of winning the big prize.

Chapter 4

Federal Statistics

The federal government provides a constant stream of information of interest to the public, from inflation figures to unemployment numbers.

Reporters should understand where these numbers come from and how they can be used.

This section examines the unemployment rate, inflation (as indicated by the Consumer Price Index), the gross domestic product, and the international trade balance.

Current unemployment and CPI figures are found on the website of the Bureau of Labor Statistics, www.bls.gov. GDP and trade balance figures are found on the site of the Bureau of Economic Analysis, www.bea.gov.

Numbers in the news...

Prices Up

Consumer prices rose modestly in October, especially hitting the wallets of motorists and Americans needing medical care. But falling prices for airfares, computers and other items provided some bargains.

The government's most closely watched inflation gauge, the Consumer Price Index, increased 0.3 percent in October from the previous month, the Labor Department reported Tuesday.

The latest reading on inflation...followed a 0.2 percent advance in September and equaled the rise posted in August.

Associated Press, Nov. 20, 2002

4.1 Unemployment

Every month the U.S. Department of Labor, Bureau of Labor Statistics issues a report on the employment situation in the United States. The unemployment rate is defined as the percentage of the labor force that is unemployed and actively seeking work.

But what are the facts behind these figures? In order to get the whole picture, it's important to understand how the Department of Labor comes up with its numbers.

First, let's examine the term *labor force*. It means everybody over the age of 16 who has a job or has looked for one in the past four weeks, except unemployed people who are not actively seeking work (either because they've given up trying to find a job or they are retired, a stay-at-home parent, etc.) and people who are institutionalized (such as in prison). In 2000, the U.S. labor force averaged about 141 million people.

Being *employed* means the person did some work (even one hour) for pay in the week before the survey was taken, or did at least 15 hours of unpaid work for a family enterprise. If the person was on vacation, on strike, or away from work for some other reason but has a job, that person also is counted as employed.

Obviously, the government doesn't survey every single person in the country and ask if he or she is working. Instead, a group of 60,000 households, called the Current Population Survey, is interviewed once each month. The respondents are asked about their work habits for the previous month. Data from this survey are used to create the unemployment figures for each state and the nation.

The BLS adjusts some statistics to take into account seasonal employment changes. For example, around Christmas employment suddenly goes up as the retail

---

Formula
Unemployment rate = (unemployed ÷ labor force) x 100

---

market heats up. The BLS adjusts the figures so that December employment numbers don't seem unusually better than other months. Annual figures, however, are not adjusted.

For more information on how the unemployment rate is calculated, visit the Bureau of Labor Statistics website at www.bls.gov.

4.2 Inflation and Consumer Price Index

Inflation is an issue journalists frequently face. Some years it's worse than others, but inflation in some form almost continuously affects the economy.

U.S. inflation is measured by the Consumer Price Index. This figure, determined by the Bureau of Labor Statistics, shows the amount of inflation in any given month for eight major product groups, including food and beverages, housing, apparel, transportation

Unemployment story ideas

■ Remember that the Bureau of Labor Statistics considers somebody to be employed even if they only work part time. Are the part-time workers in your community working part time by choice, or because they can't find a full-time job? What does this tell you about the employment statistics for your community?

■ "Discouraged workers" are those who could work, but have stopped searching. They are not counted as part of the official labor force, so they're not counted as unemployed. Statistics show that the official labor force goes up when unemployment goes down, possibly because many discouraged workers start looking for work again when there are more jobs available. Does the number of discouraged workers throw off the unemployment figures in your community? Why are they getting discouraged?

and recreation.

The items included in the index were determined by a series of surveys on the spending habits of about 30,000 families and individuals.

CPI data are collected by BLS employees who visit or call about 23,000 retail and service businesses each month. They note the prices of items that were precisely defined in an earlier visit. Information on rents is collected from about 50,000 landlords and tenants. All this information is sent to the BLS office in Washington, D.C., and used to compute the Consumer Price Index.

The CPI is reported in several ways. Sometimes it is written as an *index number*, which is some number more than 100. In these cases, the figure shows how much prices have increased since the base CPI, 100, was created in 1984. For example, the CPI in September 2002 was 180.8. This means prices for the tracked items increased 80.8 percent from 1984 to September 2002.

Other times the change in the CPI is reported as a monthly or annual inflation rate. The Bureau of Labor Statistics website lists the current inflation rate as a percentage, but it's good to know the math behind those figures.

Formula

Monthly Inflation Rate =
(Current CPI - Prior Month CPI) ÷ Prior Month CPI x 100

Example

The August 2002 CPI was 180.7. The September 2002 CPI was 181.

Inflation rate for September 2002 = (181 − 180.7) ÷ 180.7 x 100 = 0.166%

This is rounded to the nearest tenth of a percent, so the inflation rate for September 2002 was 0.2%

The BLS also reports an annual inflation rate each month. This is determined by comparing the current CPI with the CPI of that month in the previous year.

Formula

A = Annual Inflation Rate
B = Current month CPI
C = CPI from same month in previous year

A = (B - C) ÷ C x 100

Example

The August 2002 CPI was 180.7. The August 2001 CPI was 177.5.

Annual inflation rate = (180.7 - 177.5) ÷ 177.5 x 100 = 1.8%

So the annual inflation rate as measured in August 2002 was 1.8%

Adjusting for inflation

You have probably heard the phrase "adjusted for inflation." This means a historical figure was changed to represent how large it would be in current dollars. You can do this using historical CPI information available from the Bureau of Labor Statistics website. The website makes it easy to calculate this figure with an "inflation calculator" that allows users to type in the dollar amount of the figure in question and the year. The math behind that calculator is:

Formula
A = Target year value, in dollars
B = Starting year value, in dollars
AC = Target year CPI
BC = Starting year CPI

A = (B ÷ BC) x AC

Example

When Rick Strawhecker started out as a journalist in 1971 he was paid $6,500 a year at the Eubank Eagle. How much would this be in "2002 dollars"?

You can do this either by using the calculator on the BLS website, or by finding the CPI for 1971 and the CPI for 2002 and using the above formula. You can find the average annual CPI on the far right of the historical CPI table on the BLS website.

2002 Salary = ($6,500 ÷ 40.5) x 179.9 = $28,872.84

So Strawhecker's 1971 salary is the equivalent of a $28,872.84 salary in 2002.

What if you wanted to figure out how much something would cost a year from now? The CPI doesn't predict future prices, but if you know the current rate of increase of the CPI, you can calculate what an item will cost if that rate were to remain precisely the same by finding the annual rate and applying it to the original price. As you learned in the chapter on percentages, it is more accurate to compound an interest rate when it is being applied over a long period of time. For the purpose of applying inflation, we'll compound monthly.

Formula

C = cost after one year
K = original cost
I = inflation rate

$$C = K(1 + [I \div 12])^{12}$$

Example

The annual rate of inflation measured in January 2003 was 2.6 percent. Let's say a typical family of four spends $350 per week on food, clothing and other household goods. If the inflation rate remained the same, how much would the typical family pay for those items in January 2004? Remember that to use a percent in a calculation you need to convert it to a decimal by dividing by 100; so 2.6 percent becomes .026.

2004 Cost = $350(1 + [.026 ÷ 12])^{12} = $359.21

4.3 Gross Domestic Product

Gross Domestic Product (GDP) is the value of goods and services produced by a nation's economy. It is a handy number to gauge the direction of the country's economic activity. When GDP is increasing, the economy is considered healthy, and when GDP is decreasing, the economy may be in a recession. It is the *change* in GDP, rather than its level, that is most closely watched.

The GDP is often converted into "real" GDP, which holds the prices of the measured items constant to the prices they were in 1996. Real GDP, thus, shows changes in quantities of goods and services produced over the years, without being affected by price changes.

GDP is reported quarterly, and the rate of GDP growth is reported as an annual rate. For example, the real GDP during the fourth quarter of 2002 increased at an annual rate 1.4 percent. This means if the GDP increased at the rate it did in the fourth quarter, it would be 1.4 percent higher at the end of the year than it was at the beginning.

Formula

C = consumer spending on goods and services
I = investment spending
G = government spending
NX = net exports (exports minus imports)

GDP = C + I + G + NX

Another figure is real GDP per capita, which measures goods and services produced per person. This measure can be used to compare the relative well-being of populations in different countries. For example, mainland China's GDP in 2001 was $1.16 trillion, while Hong Kong's was only $161 billion. But because China's population is so large, its per capita GDP was only $910, compared with nearly $24,000 for Hong Kong. Thus, in terms of goods and services produced per person, Hong Kong's population is better off than China's.

The U.S. Bureau of Economic Analysis collects the data for the Gross Domestic Product from a wide range of sources. These include other government agencies such as the Bureau of Labor Statistics and the Bureau of Census, and private companies such as Ward's Automotive Reports and R.L. Polk & Company.

4.4 Trade balance

One often reads about the trade balance. The trade balance is the difference between the goods and services a country exports to foreign countries and those it imports. For the United States this has been a negative number for years, which means Americans are importing more goods than they are exporting.

There are seven major categories of exports and

imports. The largest category for the United States in both exports and imports is capital goods other than autos (this category includes aircraft, computer accessories, machinery, etc.). The second largest export is services, which includes travel (lodging, recreation, passenger fares), royalties and license fees, and other private services. The other five categories are industrial supplies, autos and auto parts, consumer goods, food and beverages, and "other."

The data for trade balance are collected from a large number of sources, and are considered comprehensive (that is, they are a complete count of activity; they're not extrapolated from sample data). The most important source is the U.S. Customs Service, which collects detailed information on goods exported and imported.

Formula
Trade balance = Exports - imports

Example
In January 2003, America exported $81.9 billion worth of goods, and imported $123 billion.

Trade balance = $81.9 billion - $123 billion = -$41.1 billion

The trade balance for January 2003 was negative $41.1 billion. Thus, it was a trade deficit.

Skill Drills

1. The CPI in July 1990 was 130.4. It increased to 131.6 by August 1990. What was the percentage increase in CPI for that month?

2. In 1976 the mayor of Jonesville was paid $22,000. The CPI for that year averaged 57. In 2002, when the CPI averaged 179, the mayor of Jonesville earned $56,000. Accounting for inflation, which mayor earned more?

3. Say the current annual rate of increase of the CPI is 2.4 percent. If a grocery bag of common household items costs $95 today, how much will it cost one year from now?

4. In 1996 American companies exported $30.8 billion worth of aircraft components and imported $12.7 billion worth. What was the trade balance in aircraft components?

Learning Challenge

Ask your parents what they paid for tuition, their first car and their first house. Use the CPI index to calculate and compare similar values for today. Are you keeping up with inflation, or falling behind?

Chapter 5

Polls and Surveys

Polls and surveys are as much a part of the American landscape as elections and Fourth of July parades.

From candidates for political office testing their support to marketers seeking the views of consumers, polls and surveys are used, reused, reported and reviewed. However, polls and surveys offer only a glimpse of public opinion, and often are skewed one way or another. It is the reporter's job to help readers understand the validity of polls and surveys they are reading about.

First some definitions. Polls are an estimate of public opinion on a single topic or question. Polls are most frequently used in

Numbers in the news...

Poll Questioned

When a conference committee of Tennessee legislators convenes tomorrow to hammer out a balanced budget, proponents of one plan are likely to roll out a recent poll they say shows a high level of support statewide for a graduated income tax...

The poll of 3,000 registered voters statewide indicates that 62 percent support the tax.

A professor contends that the large sample size gives his telephone poll a margin of error of only 1.8 percentage points.

The Tennessean, Nashville, Tenn., May 28, 2001

political circles and are based on representative samples of the population. Surveys are also based on representative samples of the population but usually include multiple questions and are used in a wide variety of social science settings.

One important issue to consider in evaluating a poll for reliability is the concept of random selection. Random selection means that every person in the population being studied has an equal opportunity to be selected. That eliminates surveys in which respondents select themselves for participation (such as web surveys that can be answered by anybody who happens to come across the website).

In this chapter, we'll discuss the formulas pollsters use for selecting samples, what margins of error and confidence levels are and how they're calculated, how to use the U.S. Census, and what z and t scores are.

5.1 Populations and samples

Ideally, researchers would like to interview everyone in the designated population to get feedback on the issue under study. However, that is seldom possible. Instead, researchers rely on samples to generate a picture of the population.

A valid sample is large enough to represent the population under study. Pollsters aim for at least 400 interviews to keep the margin of error within acceptable limits.

Formulas for selecting samples

■ Census, universe or population sampling involves sampling everyone in the population. The U.S. Census is a population sample because every household is queried. Population samples are too expensive, however, to be conducted on a regular basis.

■ Cluster sampling involves sampling in one area or region identified by ZIP codes, counties or other well-defined areas. For example, a sample of students enrolled in Media Writing could be a cluster sample of journalism majors.

■ Multistage sampling is frequently used in national samples. This process involves selecting a specific geographic area, then randomly selecting sub-groups, then individual blocks within the sub-group and finally a smaller block.

■ Systematic random sampling calls for selecting a specific number, say 10, and using the phone book, city directory or other reference book and polling every 10th person.

■ Quota sampling aims to select the sample based on known demographic characteristics. For example, in a poll on women with school-age children who work outside the home, a sample could be devised that would include a proportionate number of women who work in offices and women who work in factories.

■ Probability sampling involves putting all of the potential subjects in a hat and drawing out a designated percentage. The method is time-consuming and costly, but also relatively reliable because everyone in the population has an equal chance of selection.

5.2 Margin of error and confidence level

Margin of error and confidence level are key terms and concepts needed to understand and report accurately on polls, surveys, and scientific and academic studies.

Margin of error

Margin of error indicates the degree of accuracy of the research based on standard norms. Statisticians concerned with reducing mistakes, misinterpretations and flaws in procedures developed the figures, formulas and procedures.

Margin of error is expressed as a percentage and is based on the size of a randomly selected sample. Statisticians have developed specific percentages based on specific sample sizes. The more people polled, the smaller the possible error, thus the smaller the margin of error.

Statisticians have generated charts showing the margin of error based on sample size (see sidebar on page 73).

Example

Reporter Latisha DeWine received a press release from the Center for Candidate Polling. The press release indicated that the center conducted a random poll of 500 female voters following a debate between candidate Cecil Lax and candidate Marybeth Ludwig on their positions on several specific issues of interest to women. Lax received support from 53 percent of those polled. Ludwig received support from 47 percent.

Would it be correct to say that Lax was leading in the poll?

No.

The margin of error for a poll of 500 people at the 95 percent confidence level is 4.4 percent (see the note about percent vs. percentage point below). That means the support for Lax could range between 57.4 percent

Margin of error for different sample sizes

Sample size	95% confidence level	99% confidence level
50	13.9	18.2
100	11.3	12.9
200	6.9	9.1
300	5.7	7.4
400	4.9	6.4
500	4.4	5.7
1,000	3.1	4.1
5,000	1.4	1.8

Notice that the margin of error increases as the confidence level increases.

(53 + 4.4 = 57.4) and 48.6 percent (53 - 4.4 = 48.6). The support for Ludwig could range between 51.4 percent (47 + 4.4 = 51.4) and 42.6 percent (47 - 4.4 = 42.6). The two percentages overlap (Lax 48.6-57.4, Ludwig 42.6-51.4), indicating that there is little statistical difference in the candidates' level of support.

The margin of error should be included in all stories related to polling along with an accurate interpretation of the meaning of the percentage.

■ **Note: Percent vs. Percentage Point in Margin of Error**

In chapter 2 you learned the difference between percent and percentage point, and that they are not normally interchangeable. Margins of error are given as percents, but when you add or subtract the margin of error from the result of a poll, it is treated as if it were expressed in percentage points (that is, you simply add or subtract the margin of error figure from the result; you do not calculate the percentage). Why? Because the margin of error refers to a percentage of the actual polled number, not a percentage of the result. In the example above, of the 500 people polled, 53 percent preferred candidate Lax. This means 265 people preferred Lax, and 235 preferred Ludwig. The margin of

Election-speak

■ Electorate: the entire population of a district eligible to vote.

■ Exit poll: a poll in which voters leaving a polling station are asked how they voted.

■ Majority: More than 50 percent of votes cast.

■ Percentage of precincts reporting: Determined by dividing the number of precincts that have reported results by the total number of precincts.

■ Plurality: When a candidate receives more votes than the opponents, but less than 50 percent.

■ Straw poll: A nonscientific poll.

■ Turnout: A percentage based on the number of registered voters in the district and the number of voters who cast ballots. Determined by dividing the number of voters by the total number of registered voters.

■ Winning margin: The difference between the highest number or percentage of votes cast and the next highest number of votes cast or percentage cast.

error was 4.4 percent of 500 people (22), not 4.4 percent of 53 percent. Add 22 to 265 and you get 287, which is 57.4 percent of 500. Subtract 22 from 265 and you get 243, or 48.6 percent. This is precisely what you would get if you added or subtracted the margin of error (4.4 percent) from the result (53 percent, in Lax's case).

Confidence level

Another issue in reporting polls is the confidence level, which is a bit more complicated to understand, but goes to the heart of accurate reporting.

A simple explanation of confidence level is that it is the level, or percentage, at which researchers have confidence in the results of their research. The formal definition is the probability of obtaining a given result by chance.

The confidence level is determined in

advance and usually falls at 90 percent, 95 percent or 98 percent. Different fields require different confidence levels. A confidence level of 90 percent means the research results had a 10 percent probability of occurring by chance. Researchers select the confidence level in advance based on pre-testing and previous research.

The confidence level figure is closely related to the margin of error. The confidence level should always be reported as part of the story because it gives the readers a chance to assess the results for themselves.

Reporters seldom have to calculate the margin of error, but some degree of familiarity is called for to assess the accuracy of test results. This helps readers grasp the significance, or lack thereof, of the statistics.

Example

In the margin of error example above, the results were reported at the 95 percent confidence level. However, if the researchers had wanted a 99 percent confidence level, the margin of error would have been 5.7 percent (as the confidence level goes up, the margin of error gets larger).

At the 99 percent confidence level, Lax's percentage of support ranged from 58.7 percent (53 + 5.7) to 47.3 percent (53 - 5.7). Ludwig's support ranged from 52.7 percent (47 + 5.7) to 41.3 percent (47 - 5.7). The percentages still overlapped. There was no significant difference in the poll results.

To report the results, you would write:

Female voters polled after the debate indicated that 53 percent supported Lax and 47 percent supported Ludwig. Because the poll was based on responses from 500 people there was a margin of error of plus or minus 4.4 percent, indicating there was no significant difference in the candidates' support after the debate. The margin of error is based on a 95 percent confidence level, which means there is a 5 percent probability the results were the result of chance.

5.3 Census

The 2000 U.S. Census was a mail and door-knocking survey of every household in the United States. Results are slowly trickling out and will continue for several years. The return rate was 67 percent.

Short forms were sent to 5/6 of households. The 1/6 of households that received the long forms were asked the same basic seven questions as the households sent the short forms, and 52 additional questions on social, economic, financial, and housing information.

Adjusted versus unadjusted figures

Adjusted figures are figures that are statistically manipulated to compensate for missing data. Researchers have determined that an undercount tends to occur in predominately black areas while overcounts tend to occur in predominately white districts.

The results of the Census are used for creating congressional districts relatively equal in population. The districts are based on unadjusted figures following a 1999 Supreme Court ruling that adjusted figures could not be used.

However, the formula used is designed to give smaller states slightly more representation. For example, Delaware needs fewer new residents than California to pick up an additional seat in Congress.

The states also use the results to recreate legislative districts. The states may use adjusted figures. The Justice Department reserves the right to ask a state to justify using adjusted versus unadjusted figures.

The U.S. Census releases adjusted and unadjusted numbers for every level. It is expected that on the state levels adjusted and unadjusted figures will vary from 1 percent to 4 percent, depending on the issue.

Using the numbers

Local city planners have copies of official census tracts in the form of maps. Census tracts seldom

---

change. They are controlled by the U.S. Census.

Block comparisons are valuable because they are based on 100 percent data. Analysis of multiple blocks also generates valid information about a community.

5.4 z scores and t scores

Journalists should have a basic understanding of z and t scores because they are often used in reporting the results of studies. Knowing how these scores are calculated and used will enable a reporter to better assess information provided by researchers.

A z score, also sometimes called a "standard score," shows how much a particular figure differs from the mean. It is commonly used to compare figures that are hard to compare in their raw form.

In a z score, the standard deviation (see page 42) is used as the unit measure. The mean becomes zero and the first standard deviation is 1, the second is 2 and so forth. Z scores can be negative or positive depending on the location relative to the mean.

The t score, also sometimes called Student's t distribution, is closely related to z scores. It is used when the sample size is small, roughly 100 or fewer. This calculation is beyond the scope of this book and it requires a table of "critical" t values; consult a statistics text if you need to calculate t score.

Formula

z score = (Raw score - mean) ÷ standard deviation

Skill Drills

1. Using the campus phone directory identify every 9th person listed under the letter W. How many people did you count? Define the sampling method.

2. The Graduate Research Methods class decided to find out how many students on campus read a newspaper at least five times a week. Select one of the five sampling methods. Justify your selection.

3. Use the chart in the box on page 73 to answer questions 3a-3c below. The chart is based on a standard margin of error sampling chart. Should you have a sample size that falls between the stated numbers, select the sample size closest to your sample to get a viable figure.

3a. Chen Zhang conducted a study of the types of sources broadcast reporters use for their stories. She obtained data on 500 stories, randomly selected. Zhang found that 22.7 percent of the stories were suggested by reporters, news managers suggested 43.2 percent of the story ideas, 15.9 percent came from network feeds and local newspapers accounted for 9.6 percent. Zheng was unable to determine the source of 8.6 percent of the stories. What was the margin of error for her study at the 95 percent confidence level? What was the margin of error at the 99 percent confidence level? Were any of the results significant?

3b. Trina Wellington is the education reporter for the Valley Vision. Wellington polled the parents of students enrolled in the Valley Public Schools regarding their views on public education. She received 1,000 responses and found that 68 percent said the schools were doing an excellent job, 24 percent said the schools were doing an average job and the remainder were extremely critical of the schools. If she used a 95 percent confidence level, what was the margin of error?

3c. Using the sample chart, calculate where the greatest differences occur in margin of error between adjoining numbers using numbers from the 95 percent confidence column. Do the numbers decrease in any pattern? Where is the smallest difference? Where is the largest difference? What conclusion can you draw regarding the differences?

4. Go to the U.S. Census website (www.census.gov) to complete the following assignments.

4a. What was the April 1, 2000, population of your hometown, county and state?

4b. Did the population increase or decrease in your hometown, county or state? By what percent?

4c. Find the street address finder on the census page. Plug in your street address. Print out the map showing your neighborhood. Can you find any basic demographic information about your neighborhood on the census web site?

4d. What is the racial and ethnic background of your hometown, county and state? How do these numbers compare to national figures? How do these figures compare to figures released following the 1990 census?

What conclusions can you draw when looking at the figures? Write a summary lead on the changes focusing on either your city, county or state?

Learning Challenge

The U.S. Census web site, www.census.gov, contains much demographic information that can be used to flesh out a variety of stories, from crime stories to business stories. The site also provides detailed community profiles that can enhance reporting on almost any topic in the news.

1. Use the U.S. Census web site to find out information about specific census blocks in your community. What is the gender, racial/ethnic character? What information is available regarding housing, business and other economic characteristics? Locate information from 100 years ago and compare the information to the most recent information. If your campus library has a federal depository library, additional census information should be available in the federal collections.

2. Survey journalism students regarding their media reading habits. Include newspapers, magazines, television and online news sources. Consider, discuss and debate the pros and cons of using population sampling, cluster, multistage sampling, systematic random sampling, quota and probability sampling. Use inferential statistics to put the scores of sub-groups (freshman, sophomores, juniors, seniors or print versus broadcast) into perspective.

Chapter 6

Business

Business news is often big news, and few beats contain as much math as the business beat. There are many sources of business news — press releases, quarterly earnings reports, annual reports — but it's important to understand the basics before reporting from these sources.

Large corporations usually report earnings results quarterly. Quarterly financials are used by investors, banks, brokers and others to assess the financial condition of a company.

More detailed information can be found in annual reports. Annual reports are readily available on-line, through stockbrokers and in some libraries. In addition to investors and others interested in a company's

Numbers in the news...

Truckin' Along

A sudden rise in orders has the troubled truck industry talking about a mini rebound. But Navistar International Corp. is keeping its excitement in check...

"We won't ramp up our own production for what we believe to be a short-term blip in sales," said Navistar president J. Steven Keate.

In its first quarter, Navistar's revenues slipped 3 percent to $1.47 billion and the company recorded a loss of $56 million, or 93 cents per share, vs. a year-earlier loss of $35 million, or 58 cents a share.

Crain's Chicago Business, Feb. 25, 2002

financial situation, annual reports are read by the Internal Revenue Service to review documents relating to a company's tax obligations.

6.1 Financial statements

Financial statements are formal documents available to shareholders, regulatory agencies, and other stakeholders interested in a company's performance. The documents include quantitative statements about a company's business transactions.

Financial statements, which are found in a company's annual report, come in many different forms, but they generally include some type of profit and loss report and a balance sheet.

6.2 Profit and loss

The profit and loss statement, commonly called P&L, is one of the most important documents a company issues. It shows whether a company is making money or not.

The Coca-Cola Company published a rather concise P&L in its 2001 annual report. It is reproduced at right.

Different methods are used by different businesses to determine profit and loss. Basically, though, every business calculates profit by subtracting expenses from income.

A business has many different expenses. The phrase "cost of goods sold" refers to the direct expense a business incurs in making or buying its products. For example, Frank's Furniture Factory, which makes dining room tables, would consider its wood, nails, glue, and varnish as its "cost of goods sold." If a company buys finished products for resale, the cost of goods sold is often simply called the "wholesale" cost.

The word "overhead" refers to the expenses not directly related to the product being made, including salaries of employees, rent, utilities, insurance, etc.

2001 Coca-Cola Consolidated Statements of Income (P&L) (in millions)

Net operating revenue...........................$20,092
(How much money the company earned from selling Coca-Cola and other products)

Cost of goods sold......................................6,044
(How much it cost the company to make that Coca-Cola and other products, such as the cost of sugar and water)

Gross profit 14,048
(How much was left after the cost of goods sold was subtracted from the net revenue)

Selling, admin and general expenses 8,696
(This includes administrative salaries, rents, office supplies and thousands of other expenses)

Operating income 5,352
(Gross profit minus the selling and admin expenses)

Interest income...325
Interest expense..(289)
Equity income...152
Other income 39
Gains on issuance of stock 91
(Other sources of income and expenses. Note that *interest expense* will be subtracted while the other items will be added to income)

Income before income taxes and effect of accounting change...5,670

Income taxes..1,691
Cumulative effect of accounting change ......(10)
(Coca-Cola accountants changed some procedure sometime during the year, resulting in this 'paper' change to the financials. This amount is subtracted [that's why it's in parentheses])

Net income 3,969
(The *bottom line*. Remember it's in millions)

Financial statement tips

■ Numbers are often written in "thousands" or "millions," i.e., the last three or six zeroes are deleted. This is disclosed at the top of the financial statements, so look for it.

■ Figures in parentheses are negative.

■ Compare net income figures over time. It is more newsworthy to report a difference over time or a trend than to simply report one year's numbers.

The difference between the "cost of goods sold" and the selling price is considered the "gross margin." For example, let's say the direct cost for Frank's Furniture Factory to make one dining room table is $100 (the cost of the wood, nails, glue, and varnish). If Frank's sells that table for $250, its gross margin is $150. Sometimes, particularly in retail settings, gross margin is called "mark up."

Now let's say Frank's sells 100 such tables in a month. Its gross margin on those sales is 100 x $150, or $15,000. Is that the company's profit for the month? No. All of the other company's expenses, its overhead, also need to be subtracted. Let's say the company's payroll for the month is $5,000, its rent is $1,000, utilities are $500, its marketing cost is $1,500, and other overhead items are $2,000. So the total overhead for the month is all those items added together, $10,000.

Subtract $10,000 from $15,000 and you get $5,000. Even that won't go entirely into the owner's pocket. He may have to pay taxes on that, and he may have to pay down loans. And some of his equipment may have depreciated over the month, so that amount gets subtracted from his profit (on paper for tax purposes).

When all these subtractions have been made, the owner is left with the "net profit." This is sometimes called "net earnings" or "net income."

Sometimes it is useful to see how a company is doing before certain expenses, such as interest pay-

ments and deprecia-
tion, are subtracted.
For example, a bank
considering a loan
may want to see how
much cash a company
actually generates in a
month before depreci-
ation and amortization
are figured in (depreci-
ation and amortization
are figures that meas-
ure how much value
certain assets have
lost; they are "paper"
expenses, because the

company doesn't really lose cash when its assets lose
value).

In these cases, it is useful to calculate EBITDA,
which stands for "earnings before interest, taxes, depreciation and amortization."

EBITDA also is a useful figure to compare companies, because it shows how much cash a company is
earning without regard to items unrelated to current
business (interest payments, taxes, depreciation, and
amortization need to be accounted for no matter how a
company is doing).

EBITDA is often called "operational cash flow,"
because it measures how much cash the business has to
operate with.

Formula
Gross margin = Selling price - cost of goods sold

Example
Will Raines worked part-time for the Reynolds
Review selling copies of the Sunday newspaper from a
newsstand on the Circle downtown. Raines paid 90

cents for each copy and sold them $1.25. What was his gross margin?

1.25 - .90 = 35 cents

Formula
Gross profit = Gross margin x number of items sold

Example
In the example above we learned that Will Raines has a margin of 35 cents per Sunday paper he sells. If he sells 50, what is his gross profit?

35 cents x 50 = $17.50

Formula
Net profit = gross margin - overhead

Example
Will Raines pays $5 to rent a newsstand on Sunday mornings. This is his only overhead expense. What is his net profit for an average Sunday, using his gross profit figures from above?

$17.50 - $5 = $12.50

6.3 Balance sheet
A balance sheet is a written financial statement of a company's assets, liabilities, and equity. It is another important business document, because it shows the financial stability of the company.

As with the profit and loss statement, different companies use different terms in their balance sheets.

The common denominator of all balance sheets is

that the assets side of the balance sheet always equals the liabilities and equity side.

Formula

Assets = Liabilities + Equity

This may seem counterintuitive. Why would liabilities plus equity equal assets? This concept is easier to understand if you turn the equation around: Assets - Liabilities = Equity. This means that the company's assets (such as its real estate, equipment, cash, etc.) minus the liabilities (the money the company owes) equals the equity (essentially what the company is really worth).

Example

In 2001 the Coca-Cola Company reported $11,051,000,000 in liabilities and $11,366,000,000 in equity.

$11,051 + $11,366= $22,417 in assets (in millions)

Balance sheet definitions

The 2001 balance sheet from The Coca-Cola Company is reproduced (with some modifications for clarity) on the following pages. The balance sheet has definitions for terms written under most items, but here are general definitions of terms found on most balance sheets:

■ *Assets* are resources owned by a company that have some economic value. *Current assets* include cash, investments, and other liquid items of value. *Long-term assets* include buildings, office furniture, etc.

■ *Accounts receivable* represents money owed to the company by customers.

■ *Accumulated depreciation* is the decline of the value of

2001 Consolidated Balance Sheet (In millions)
The Coca-Cola Company

Assets

Current

Cash and cash equivalents..............................$1,866
Marketable securities..68
(Investments that could be easily turned into cash)

Accounts receivable 1,882
(Money other companies owe Coca-Cola)

Inventories..1,055
(Warehouses full of sugar, caramel coloring, etc.)

Prepaid expenses and other assets 2,300

Total current assets..7,171

Investments and other assets

Coca-Cola Enterprises Inc. 788
Coca-Cola Amatil Limited..432
Coca-Cola HBC S.A. 791
Other, principally bottling companies 3,411
(The four items above are investments Coca-Cola has made in
its related companies)

Other investment assets 2,792

Total investment assets.......................................8,214

Property, Plant and Equipment

Land 217
Buildings and improvements................................1,812
Machinery and equipment....................................4,881
Containers 195

Total property, plant and equipment 7,105

Less allowances for depreciation 2,652
(The amount of value the assets have lost due to usual wear and
tear. This is subtracted from the assets.)

Trademarks and other intangible assets..............2,579
(The Coca-Cola trademark is worth a lot of money)

Total Assets 22,417

Liabilities and Share-Owners' Equity
Current
Accounts payable and accrued expenses 3,679
(What Coca-Cola owes to suppliers and others)

Loans and notes payable 3,743
Current maturities of long-term debt 156
(The two items above are loans, or portions of loans, the company has to pay back within one year)

Accrued income taxes 851
(Taxes that are owed, but not yet paid)

Total current liabilities 8,429

Long-term debt 1,219
(Debt that eventually needs to be repaid, but not within a year)

Other liabilities 961

Deferred income taxes..442
(Taxes that will eventually have to be paid, but not in the short term. These generally come from tax-deferred investments)

Share-owners' equity
Common stock and capital surplus......................4,393
(Coca-Cola stock owned by shareholders and other capital)

Reinvested earnings 20,655
(Cash kept in the company)

Less treasury stock...(13,682)
(Coca-Cola stock the company bought back from shareholders. This amount is subtracted from the total)

Total Share-owners' equity 11.366

Total Liabilities 22,417
(Notice that this figure equals Total Assets)

an asset. For example, a car depreciates in value as it is driven.

■ *Inventory* is a record of goods on hand, raw materials, and work in progress.

■ *Intangible assets* include copyrights, patents, and research with legal and economic value.

■ *Investments in other corporations* can be in the form of stock owned or influence acquired in other companies for economic gain.

■ *Fixed assets* include property, plants, equipment, and deferred charges.

■ *Short-term investments* include stocks and bonds.

■ *Pre-paid expenses* include rent and insurance.

■ *Uncollected accounts receivable* includes write-offs for bad debts and an allowance for potential bad debts.

■ *Equity* means the value of the company, and normally refers to the owners' and/or shareholders' investments in the company, capital accounts, and other related assets.

■ *Dividends* are payments to shareholders that represent the distribution of the company's assets.

■ *Retained earnings* are earnings set aside for future business purposes.

■ *Liabilities* are obligations, such as loans, that need to be paid at some later date.

■ *Accounts payable* are bills that need to be paid.

■ *Accrued liabilities* are liabilities that have occurred but are not yet paid.

■ *Current* or *short-term liabilities* includes money owed to suppliers, interest on debt, taxes, and wages.

■ *Long-term liabilities* include debt, deferred taxes, and leases.

6.4 Ratio analysis

Ratios are calculations that analysts and business owners use to evaluate a company's cash situation, profitability, operating efficiency, and market value. They examine trends in a company's life, and are often used to compare a company against other companies in the same field.

When reporting a company's ratios, remember that they are only indicators of a company's strengths or weaknesses and should be recognized for their limitations. You should try to include the industry standard ratios when reporting a company's ratio, so your readers can compare. Industry standards can be found in Dun & Bradstreet's Industry Norms and Key Business Ratios, a publication probably located in your library.

Current ratio

Current ratio is a liquidity ratio that measures the ability of a company to meet its liabilities. It is one of the ratios you will see most often.

Formula
Current ratio = current assets ÷ current liabilities

Example

Pickens Media Co. has $104.8 million in current assets and $86.6 million in current liabilities.

104,808,000 ÷ 86,600,000 = 1.21

A ratio of 1.21 means Pickens Media Co. has $1.21 in assets for each dollar in liabilities.

What does this number mean to you? Probably nothing, unless you know that the average current ratio for media companies is 1.8. By comparing, you can see that Pickens Media Co. has a current ratio that is below industry standards. That means it may not be able to pay its bills as well as another media company with a higher current ratio.

Quick ratio

Quick ratio is a liquidity ratio that measures the ability of a company to meet its current liabilities with cash on hand.

Formula
Quick ratio = cash ÷ current liabilities

Example
Of the $104.8 million in current assets that Pickens Media holds, $24 million is cash.

$24,000,000 ÷ $86,600,000 = .277

This means that Pickens has only about 28 cents in cash for each dollar in liabilities it faces. Is that bad? Compared to the industry standard of 1.3, i.e. $1.30 cash for each $1 of liabilities, Pickens is short on cash.

Debt-to-asset ratio

Debt-to-asset ratio is similar to current ratio, except it includes all assets and all liabilities (the word *debt* is often used interchangeably with *liabilities*). It is a better indicator of the long-term health of a company than current ratio.

Formula
Debt-to-asset ratio = total debt ÷ total assets

Debt-to-equity ratio

Debt-to-equity ratio tells you how deeply a company is leveraged by comparing what is owed to what is owned.

Formula
Debt-to-equity = total debt ÷ equity

Return on assets

Return on total assets is a profitability ratio that measures the return on the investment on all assets.

Formula

Return on assets = net income ÷ total assets

Return on equity

Return on equity is a profitability ratio that measures the return on the investment made in equity.

Formula

Return on equity = net income ÷ equity

Price-earnings ratio

Price-earnings ratio is a value ratio that measures the return of the investment based on stock price.

Formula

Price-earnings = market price/share ÷ earnings/share

Skill Drills

1. The town of Tuckahoe buys 1,000 tons of gravel from PK Wholesale Inc. at $10 a ton. Tuckahoe uses 670 tons to pave streets. The city manager wants to get rid of the excess gravel. He offers it to Gadsen for $13 a ton. Gadsen buys the gravel because a road collapsed and the city needs to make emergency repairs. What was Tuckahoe's gross margin on the sale?

2. PK Wholesale bought the above gravel from the gravel pit for $4.50 a ton. It was stored on PK's property for three weeks before it was sold. Dennis Morris, owner of PK, estimates his weekly overhead at $1,256 a week. This includes salaries, building expenses, truck expenses (but not the cost of the gravel). What was PK's net profit during the three weeks used for this exercise? (Assume this was the only business transaction during the three weeks.)

3. Following is a balance sheet for The New York Times. Use the balance sheet to answer the skill drill questions 3a through 3d.

The New York Times Consolidated Balance Sheets

(in thousands)	Year 2000	Year 1999
ASSETS		
CURRENT ASSETS		
Cash and cash equivalents	$69,043	$63,661
Accounts receivable	341,863	366,754
Inventories	35,064	28,650
Deferred income taxes	62,939	53,611
Assets held for sale	—	37,796
Other current assets	101,857	64,236
Total current assets	610,766	614,908
Investment in Joint Ventures	107,320	121,940

PROPERTY, PLANT AND EQUIPMENT

Land	72,228	67,149
Buildings and improvements	814,658	789,504
Equipment	1,364,256	1,307,365
Construction in progress	37,132	31,145
Total—at cost	2,288,274	2,195,163
Less accum. depreciation	1,081,114	976,767
Prop., plant and equip.—net	1,207,160	1,218,396

INTANGIBLE ASSETS ACQUIRED

Cost in excess of net assets acquired	1,060,796	953,709
Other intangible assets	419,302	351,309
Total	1,480,098	1,305,018
MISCELLANEOUS ASSETS	201,335	235,540
Total	$3,606,679	$3,495,802

LIABILITIES AND STOCKHOLDERS' EQUITY
CURRENT LIABILITIES

Comm. paper outstanding	$291,251	$———
Accounts payable	178,302	191,706
Accrued payroll and other related liabilities	114,233	105,257
Accrued expenses	203,855	193,553
Unexpired subscriptions	87,130	80,161
Curr. por. of long-term debt	2,599	102,837
Total current liabilities	877,370	673,514

OTHER LIABILITIES

Long-term debt	553,415	512,627
Capital lease obligations	83,451	85,700
Deferred income taxes	106,247	141,033
Other	705,033	634,270
Total other liabilities	1,448,146	1,373,630
Total liabilities	2,325,516	2,047,144
Total stockholders' equity	1,281,163	1,448,658
Total	$3,606,679	$3,495,802

3a. What were the New York Times Co.'s total assets in
2000? In 1999? What were the total liabilities and stockholders' equity in 2000 and 1999? Why do the numbers
match for each year?

3b. In which year(s) did The New York Times list no commercial paper outstanding and no assets for sale?

3c. Stockholders' equity represented what percent of the company's balance sheet in 1999 and 2000? What conclusion can you draw from your answer?

3d. Did The New York Times have more assets in property (plants, equipment and land) or financial assets (cash, inventory and other assets) in 2000? Was the same true for 1999?

4. Use the New York Times statements of income below to answer skill drill questions 4a through 4d.

The New York Times
Statements of Income

(in thousands)	Year 2000	Year 1999
REVENUES		
Newspapers	$3,160,247	2,857,380
Broadcast	160,297	150,130
Magazines	115,438	110,566
New York Times Digital	66,590	43,680
Intersegment eliminations	(13,117)	(5,000)
Total	3,489,455	3,156,756
OPERATING PROFIT (LOSS)		
Newspapers	$677,643	568,000
Broadcast	48,818	45,833
Magazines	19,342	18,652
New York Times Digital	(70,007)	(14,063)
Unallocated corp. expenses	(39,875)	(47,740)
Total	635,921	571,282
Income from joint ventures	15,914	17,900
Interest expense, net	64,098	50,718
Gain on dispositions of assets, and other—net	85,349	—
Income before income taxes and extraordinary item	673,086	538,464

Income taxes	275,550	228,287
NET INCOME	$397,536	310,177

4a. Compare the total revenues for 2000 and 1999. What was the percent increase?

4b. Which unit reported the highest percent increase in revenues?

4c. Compare the net income between 2000 and 1999. What was the percent increase?

4d. How much did The New York Times pay in income taxes in 2000 and 1999?

4e. Would you view New York Times Digital a profitable enterprise? Explain your answer. In which year, did NYT Digital report the smaller percentage of loss?

Learning Challenge

Obtain copies of annual reports from the top media companies. Examine fluctuations in figures, acquisitions and other business transactions over the past year. What could account for the changes?

Chapter 7

Stocks and Bonds

Stocks and bonds are two important ways businesses raise money. Bonds also are often used by governments to raise money. Understanding the basic numbers behind these cash-raising instruments is important for a journalist.

The following chapter gives an overview of stocks and bonds, credit ratings, and the role of economic indicators such as the Dow Jones Industrial Average.

For information on common ratios that analysts and and investors use to evaluate stocks, see chapter 6.

Numbers in the news...

Doughnuts Up

Mutual fund managers with cash to invest had a choice in late July. They could have paid about $28 a share for Krispy Kreme Doughnuts... or they could have paid about $45 a share for Enron Corp....

One basic argument against Krispy Kreme was that the share price then was a lofty 70 times the 40 cents a share Wall Street expected the company to earn in 2001. Enron shares were priced at 26 times expected earnings for the year. Six months later, Krispy Kreme shares are up 39 percent ...Enron, meanwhile, is all but finished as a going concern.

Los Angeles Times, Jan. 27, 2002

7.1 Stocks

Corporations sell stocks to raise cash, and people buy stocks as investments. When an individual buys a share of stock in a company, he or she becomes a part owner of the company. However, since companies generally issue millions of shares of stock, each share represents just a tiny portion of ownership.

The value of a company's stock varies over time, based on the demand. That is, the more people who want to buy a stock, the higher the price goes. The demand for a stock often goes up and down depending on business conditions, but other factors — such as publicity, future expectations, etc. — also play a role in demand.

Mutual funds are an alternative way to invest in stocks. Mutual fund companies sell shares of funds, and then use that money to buy stock in other companies. For example, an individual who wants to invest in technology companies could buy individual shares of stock in many technology companies, or he could buy shares of a mutual fund that invests in technology companies.

A typical stock table published in a newspaper looks like this:

52-week High	Low	Stock	Div	PE	Last	Change
44.80	13.30	Aon	.90	21	19.60	+.18
60.09	34.77	Apache	.40	17	54.75	-.63
51.46	40.06	Aptinv	3.28	42	43.15	-.36
26.52	16.87	Apagent		19	20.35	-.04

What does the information mean?
■ 52-week High/Low - the highest and lowest stock prices over the past year.
■ Stock - the symbol of the stock, which is usually a shortened version of the company name
■ Div - the most recent annual dividend the company paid to shareholders, per share

- PE - Price/Earnings ratio (the stock price divided by the per-share earnings reported in the last 12 months)
- Last – the price of one share at the end of the previous day
- Change – how much the stock price went up or down that day

In this example, you can see that the stock of Aon Corp. ended the day at $19.60 per share. That was up 18 cents over the day before. Sometime over the past year the stock was as high as $44.80 and as low as $13.30. The latest annual dividend the company paid was 90 cents per share, and its price to earnings ratio was 21.

7.2 Bonds

Corporations and governments raise money by selling bonds. A bond is, basically, a loan from an investor to the government or other organization selling the bond. Unlike stocks, bonds earn interest at a set rate and are generally low-risk investments.

The "face value" of the bond is the amount the owner of the bond will receive at maturity. Usually, but not always, the face value is also the amount the original buyer pays for the bond. The bond has a set interest rate, set dates for interest pay-

Investment Reports

These are some reports public companies commonly issue.

Form 8-K – Companies are required to file an 8-K when a special event occurs, such as bankruptcy, major assets are bought or sold, etc.

Form 10-K – The official audited annual report public companies are required to file. It shows assets, liabilities, revenue, etc.

Form 10-Q – Quarterly reports of important financial information.

Proxy Statement – A document sent to shareholders about matters on which the shareholders will vote.

How strong is that bond?

Credit ratings give investors an idea of the relative strength of an investment, most commonly bonds. The ratings are applied to bonds issued by corporations, governments, and other entities. They also are applied to bond funds, money market funds, and other investments.

The ratings are important to the companies issuing the bonds because the better the rating, generally the lower the interest rate they must pay on the debt.

The two largest credit rating agencies are Moody's and Standard & Poor's. These firms assign credit ratings by evaluating many factors behind the entity they are investigating, such as management, operating developments, potential regulatory action, and other factors.

The ratings for long-term investments (those maturing in more than one year) range from Aaa — which means investors are almost completely safe from any loss — to D, which indicates that the bond has defaulted in its interest payments. In between, investments rated Baa and above are termed "investment grade." Those rated Ba and below are "speculative grade" (also called "junk bonds").

The ratings themselves can be defined statistically. For example, bonds rated Aaa by Moody's average a default rate of zero, while bonds rated B have a default rate of 6.8 percent.

ments and a set maturity date. In simpler times, bonds had coupons attached to them that could be cashed like checks on interest payment dates. The term "coupon" evolved to denote the interest payment itself.

For example, say you pay $1,000 for a bond with a $1,000 face value, a 5 percent interest rate and a maturity date 30 years from now. Every year you will receive an interest check for $50, and in 30 years you'll get your $1,000 back.

The catch is, investors often sell bonds on the open market before they mature. And even though the face value and interest rate remain the same — in the case above, $1,000 and $50, respectively — the value of the bond on the open market fluctuates with supply and demand.

This means that the bond's "current yield," that is, its return on the investment, fluctuates, too. For example, the yield on the example above is exactly 5 percent when you buy the bond (if you bought it at the $1,000 face value). But if you sell the bond at $900 because you suddenly need cash and the market is down, the yield on that bond effectively goes up. Why? Because the original deal set by the bond issuer — in this case a $50 interest check each year and $1,000 pay out at the end of 30 years — remains the same no matter what price the bond was subsequently sold for. So the person who bought the bond at $900 is getting the exact same amount of money each year as the person who originally paid $1,000 for the bond — so obviously the second owner got a better deal than the original owner.

Formula
Current yield = (interest rate x face value) ÷ price

Example
Frank McKinley paid $800 for a $1,000 bond with a 6 percent interest rate. What is his current yield?

Current Yield = (6% x $1,000) ÷ $800 = 7.5%

James Washington bought the same bond a year later for $1,200. What is his current yield?

Current Yield = (6% x $1,000) ÷ $1,200 = 5%

Many bonds, particularly those issued by municipalities, are tax exempt. Thus, the return from those bonds may be slightly more than the return from another investment with the same return that isn't tax exempt, depending on the situation of the bond holder.

When considering return, remember that the calculation above doesn't take into account any reinvestment of the dividends.

Bond cost

As a reporter, you may be more interested in calculating the actual cost of a bond issued by a municipality. For example, if the park district issues $1 million in bonds to build a new pool, your readers may be interested in knowing how much those bonds will ultimately cost the park district.

Formula
Bond cost (interest) = amount x rate x years

Example

The Village of Ellsworth decided to issue $5 million worth of 10-year bonds to pay for a new fire station. If the coupon is 4 percent, how much will Ellsworth have to pay in interest over the life of the bonds?

$5 million x .04 x 10 = $2 million

If your readers are the taxpayers of Ellsworth, you may want to explain that this bond issue means $7 million in taxes ($5 million principal plus $2 million interest) will go into the new fire station.

7.3 Market indexes

Investment analysts use market averages to measure action on the exchanges. Stock indexes track the prices of certain groups of stocks, allowing investors to get a snapshot of overall market conditions without having to examine dozens of individual stock prices.

The Dow Jones Industrial Average and NASDAQ, described below, are two popular indexes. Other important stock indexes are the Russell 2000 and the S&P 500.

Bonds also are tracked by indexes, such as the J.P. Morgan Government Bond Index and the Dow Jones Corporate Bond Index.

Dow Jones Industrial Average

You've probably heard a business reporter say, "The Dow is up today...." or "the Dow fell 20 points today." He's referring to the Dow Jones Industrial Average, which is the total value of one share each of 30 select stocks divided by a figure called the *divisor*.

The divisor takes into account stock dividends, splits, spinoffs, and other applicable corporate actions (you can find the current divisor at www.cbot.com).

By monitoring the value of these key stocks, Dow Jones & Co. is able to provide a snapshot of the entire stock market.

The 30 stocks in the Dow Jones Industrial Average represent about a fifth of the $8 trillion-plus market value of all U.S. stocks and about a fourth of the value of stocks listed on the New York Stock Exchange. Dow Jones also publishes daily transportation, utility and composite indexes.

The 30 stocks in the DJIA as of January 2003 were: ALCOA, Altria Group, Inc., American Express Co., AT&T Corp., Boeing Co., Caterpillar Inc., Citigroup Inc., Coca-Cola Co., DuPont Co., Eastman Kodak Co., Exxon Mobil Corp., General Electric Co., General Motors Corp., Hewlett-Packard Co., Home Depot Inc., Honeywell International Inc., Intel Corp., International Business Machines Corp., International Paper Co., Johnson & Johnson, J.P. Morgan & Co., McDonald's Corp., Merck & Co., Microsoft Corp., Minnesota Mining & Manufacturing Co., Procter & Gamble Co., SBC Communications, United Technologies Corp., Wal-Mart Stores Inc., and Walt Disney Co.

NASDAQ

NASDAQ is an acronym for the National Association of Securities Dealers Automated Quotations. It is a subsidiary of the National Association of Securities Dealers and is monitored by the Securities and Exchange Commission.

NASDAQ is an automated quotation system that reports on trading of domestic stocks and bonds not listed on the regular stock markets. NASDAQ lists more than 5,000 securities.

To be a member of NASDAQ a company must register with the Securities and Exchange Commission, have at least two market makers (a market maker is a brokerage firm that facilitates the sale of a company's stocks and bonds) and meet minimum requirements for assets, capital, public shares, and shareholders. NASDAQ publishes two composite price indexes daily and also bank, insurance, other finance, transportation, utilities, and industrial indexes.

Skill Drills

1. Examine the newspaper stock table on page 102. Which stock's price has fluctuated the most in the past year? Which stock has the highest price to earnings ratio? Which stock price went up the most that day? Which went down the most that day?

2. If you heard about a public company going bankrupt and wanted specific details, what report might you seek out?

3. Jack Smith, business reporter for the Sundance Herald, noted that members of a local investment club paid $9,500 for a bond with a $10,000 face value and a 6 percent interest rate. What is the bond's current yield?

4. A competing club bought the same bond for $9,000 a year later. What was the current yield of that bond?

5. The Village of Augusta issued $1 million in bonds to pay for a new water tower. The term is 10 years and the coupon is 3.5 percent. How much interest will village taxpayers have to deal with over the life of the bonds?

Learning Challenge

1. Create a mock stock portfolio worth $10,000. Track the portfolio daily by noting the value of each stock. Compete with other classmates to see who can grow his or her money the fastest by "selling" and "buying" stock as you see fit.

2. Create a horizontal timeline of the leading economic indicators such as the Dow Jones and NASDAQ. Create a bar graph at the bottom tracking major news stories related to fluctuations in the economic indicators. Are there any patterns worth nothing?

3. Track the stock status of several newspaper companies over a period of time. Note fluctuations. What could be causing the stock price to rise and fall? Clip related articles from the newspapers for discussion.

4. Track the status of a mutual fund that invests in a particular category, such as technology stocks or consumer products stocks. How does its value compare to the value of individual stocks in those categories?

Chapter 8

Property Taxes

Property taxes are the largest single source of income for local government, school districts, and other municipal organizations. They pay for supplies, salaries, maintenance costs, and just about every other day-to-day expense. Articles about property taxes often make the front page, and understanding how property taxes are calculated is important for journalists.

Basically, the property tax rate is determined by taking the total amount of money the governing body needs, and dividing that among the property owners in that taxing district. How much each owner pays is based on the value of his or her property. Most taxing districts only take into account real property (i.e., real estate, homes, buildings), but some also tax cars, boats and other valuable assets.

Numbers in the news...

Budget Wrangle
The Phillipsburg Town Council introduced a $12.5 million budget yesterday...

Under the proposed budget, the town's existing level of services would be maintained and the municipal tax rate would increase from 85 cents per $100 of assessed valuation to $1.045 per $100. That is an increase of 19.5 cents per $100 assessed value.

Star-Ledger, Newark, N.J., March 9, 2001

For example, if a city determines it needs $1 million to run itself, and the 1,000 property owners in the city each own exactly the same amount of property, each would pay $1,000 in property taxes.

Of course, it's never quite that simple, but that is the underlying philosophy.

Property taxes are measured in units called *mills*. A mill is 1/10 of a cent, or $0.001. Property taxes are expressed in terms of mills levied for each dollar of assessed valuation of property. One mill per dollar works out to 10 cents per $100 assessed valuation, or $1 per $1,000 valuation.

An important thing to note is that property taxes usually are applied to assessed valuations, not to the actual price a home would sell for on the open market. The assessed value is a percentage of market value.

For example, consider the owner of a 1,600-square-foot house who received a tax assessment notice following a citywide reappraisal. The property reappraisal placed the market value of the house at $147,600. The city based its rate on an assessed value that was 25 percent of the appraised value, in this case $36,900. If the tax rate is 69.1 mills, the homeowner will pay $2,549.79 in property taxes. If the municipality assessed the property at 80 percent of true value, and kept the same tax rate, the homeowner would have paid $8,159.33 in taxes. Had the tax rate been applied to the market value of the house (the current appraisal value) the homeowner would have paid $10,199.16.

How the formula works depends on local and state government rules and regulations. Some taxing bodies include certain exemptions and other credits into the formula; others just take the straight value. Thus, it's important to learn the specifics of your community's property tax formula before writing about it.

What are some issues to consider when writing about property taxes?

A key issue is the reappraisal. The purpose of a reappraisal is to update real property values to reflect current market value of all taxable properties within a taxing district. Some neighborhoods improve in value over time and others decline, and the reappraisal process makes sure those changes are taken into account. However, how often an area is reappraised varies. Some areas reappraise property every three years, some never do.

Another key point to remember is that property is often taxed by more than one governing body. In some areas property owners pay only county taxes or only city taxes, in other areas property owners pay both. Local school boards and utility districts also levy property taxes. Thus, a city council's debate over the tax bill is only a portion of the tax picture for a property owner.

Also consider the possibility that the percentage used to calculate the assessed value might differ based on the type of property. For example, in Tennessee residential and farm real estate is assessed at 25 percent, commercial and industrial property at 40 percent and utility real estate at 55 percent. Thus, certain income-generating property owners pay a higher share of local property taxes than residential and farm property owners.

To prevent local governments from playing games with the tax rate, in many states state officials regulate the process. State laws generally provide for adjusting the tax rate to prevent local governments from benefiting from significant extra funds in reappraisal years at the expense of the taxpayers. In those states, the tax rate is never final until state authorities certify it.

This chapter provides a primer in tax rate calculations, the importance of understanding the mill levy process and how the process impacts individual homeowners.

8.1 Mill

> ## Formula
> Mill levy = Taxes to be collected by the government body ÷ assessed valuation of all property in the taxing district

Example

Kincaid's municipal budget totals $369,500 for next year. What will the tax rate be if the assessed value of all the property in Kincaid is $48,277,000?

$369,500 ÷ $48,277,000 = .00765 = 7.65 mills = $7.65 per $1,000 assessed valuation

8.2 Appraisal value

Appraisal value is based on:

■ The property's use (residential, business, vacant land, farm, business, commercial)

■ The property's characteristics including:
  ▲ Location
  ▲ Square footage (total living area)
  ▲ Number of stories
  ▲ Exterior wall type
  ▲ Age (year of construction)
  ▲ Quality of construction
  ▲ Amenities (such as the number of bathrooms and bedrooms, garage, deck, porches, fireplaces, pool, hot tub etc.)

■ Current market conditions as determined by sales in the immediate area over a specific number of years.

■ A visual inspection of the property by trained appraisers.

8.3 Assessed value

The mill levy is applied to assessed valuations. The assessed value of a property depends on local policies, which can mean credits and other adjustments. It's important to know how your municipality figures its taxes before writing about them.

Formula

Assessed value = Appraisal value x rate

Examples

Fremont assesses residential property at 25 percent of the appraised value. A house at 1949 Sweet Gum

---

Hot day, cool day?

Calculations for heating degree days and cooling degrees days are used by homeowners and electric companies to calculate the energy needed to heat or cool homes and businesses. The formula is based on 65 degrees Fahrenheit because that is the temperature where neither heat nor air conditioning is needed.

When the average temperature drops below 65 degrees Fahrenheit most buildings require heat to maintain 70 degrees Fahrenheit temperature. Each degree of mean temperature below 65 degrees Fahrenheit is counted as one heating degree-day. For each additional heating degree-day more fuel is needed to maintain 70 degrees Fahrenheit.

Formula

(High temp for the day + low temp for the day) ÷ 2 = average temp for that day

Average temp - 65 = heating/cooling degree

(If the answer is negative, it is a heating degree-day; if the number is positive it is a cooling degree-day)

Cove is appraised at $100,000. What is the assessed value of the house?

$100,000 x .25 = $25,000

Greencastle offers a credit of $25,000 if the homeowner lives in the home, and it sets the assessed value at 80 percent of appraised value. Frank Smith lives in a home at 5467 Star Light Drive that is appraised at $100,000. What is the assessed value of the house?

(Appraisal value - $25,000) x rate = assessed value

($100,000 - $25,000) x .80 = $60,000

8.4 Calculating tax

Formula

Tax owed = Tax rate x (assessed value of the property ÷ $100)

Note: Divide the assessed value by $1,000, rather than $100, if the rate is based on an amount per $1,000 of assessed value.

Example

Middletown is raising its local property tax from 75 cents per $100 (7.5 mills) to $1.032 per $100 assessed valuation (10.32 mills). How much will the property owner who owns a $125,000 house pay next year if the assessed value is based on 25 percent of the appraised value?

First calculate the assessed value:

$125,000 x .25 = $31,250

Now calculate the taxes at the old rate and the new rate:

Old rate:
(31,250/100) x 0.75 = $234.38

New rate:
(31,250/100) x 1.032 = $322.50

Now subtract the new tax amount from the old tax amount:

$322.50 - 234.38 = $88.12

The homeowner will pay $88.12 more in property taxes next year.

Skill Drills

1. Franklin's budget for the next fiscal year is $750,000. If the assessed value of all real estate in Franklin is $25 million, what is the tax rate going to be? Express that in mills and in taxes per $1,000 assessed valuation.

2. Rockham pegs its tax rate to 75 percent of true value. If a house at 5382 Brice Creek Road is appraised for $210,000, what is the assessed value of the house?

3. Kristen Lee owns a home appraised at $325,000 in Bellmead and a cabin appraised at $50,000 on the Artesia River in Andrews Township. Bellmead determines its tax rate based on 90 percent of true value while Andrews determines its tax rate based on 25 of true value. However, Bellmead provides a $20,000 credit for homeowners. What are the assessed values (less credits) of the two properties?

4. Waynesburg's tax rate for next year is $1.09 per $100 assessed value. If the house owned by Jane West has an assessed value of $56,000, how much will she pay in taxes?

5. Marion Bean considered buying houses in Millerton and Smithville. Millerton's tax rate was 45.6 cents per $100 assessed value. Smithville's tax rate was $2.28 per $1,000 assessed valuation. In which town would she pay the least in taxes for a house with an assessed value of $100,000?

Learning Challenge

To gain an understanding of the importance of tax assessment and appraisals, students have to extend their world view beyond paying sales taxes. Start with finding out how taxes are calculated in your state or community. One option is to take a trip to the assessor's office. Another option is to invite the local tax assessor to class to discuss her work.

1. Identify the 10 most expensive homes in the city/county/area. Obtain the tax records for each house. Research the histories of the homes, the builders, the neighborhoods and the owners. Do all the homes pay a proportionate amount of property tax? Does it appear that the homes were all appraised fairly?

2. Obtain the home addresses of people elected to the city council and school board. Obtain the tax assessment records for their homes, take pictures of the homes and compare them to the average cost of homes in their districts.

3. Identify the poorest neighborhood in your community using census data or other local resources. Using tax records identify owners of property in the neighborhood, the condition of the property and the value of the property. Take photographs. Identify the largest two or three property owners and research their homes. What patterns can be detected?

Chapter 9

Directional Measurements

Time, rate and distance problems are more than middle school headaches. Many news stories include directional measurements that are vital to readers.

But reporters shouldn't simply rely on numbers provided by people involved in the story. Checking the numbers in time, rate and distance problems usually involves just some basic math.

Bill Dedman, who maintains the Power Reporting site at http://power-reporting.com, has developed what has become a classic time, rate and distance problem for journalists. His question illustrates the possibility of a single number clarifying a story

Mile vs. Knot

Mile = 5,280 feet

Nautical Mile = 6,080 feet

Knot is a measure of speed. One knot is one nautical mile per hour.

for the reader.

Dedman's problem asks reporters to calculate how much time a truck driver in Bourbonnais, Ill., had to cross a set of train tracks before an Amtrak train reached the crossing. The answer is 5.6 seconds. But because the truck driver miscalculated and attempted to cross the tracks, the train hit the truck and derailed and several people were killed.

The question requires only a few steps to answer, but had reporters included it in articles about the crash, they would have answered the reader's nagging questions regarding the driver's actions that day.

The math is simple for many of the formulas in this section — just multiplication or division. But the errors generated by reporters overwhelmed by the concepts are not.

The challenge here is to be able to check the work of officials, and to do it with confidence. The result is greater accuracy for the reader or viewer, which is what is important in journalism — not to be first, but to be first with what is right.

9.1 Time, rate, and distance

Time, rate, and distance problems are also distance, rate, and time problems as well as rate, time, and distance problems. The basic formula is the same — the components get switched around depending on the solution needed.

When working time, rate, and distance problems it is important to keep the units of measurement the same. If the rate is in miles per hour (mph), then the time needs to be in hours, and the distance in miles. If the time is given in minutes, divide by 60 to convert it to hours before working the equation to find the distance in miles.

Formulas

Distance = rate x time

Rate = distance ÷ time

Time = distance ÷ rate

Examples

Time

Ernest Senn, a sportswriter for the Monmouth Mercury, was assigned to cover a bike race between Merlin and Gottlieb. Merlin is 189 miles from Gottlieb. How long would it take the average bike rider to reach Gottlieb if the rider pedaled six hours a day at an average speed of 15 mph?

Time = 189 miles ÷ 15 mph = 12.6 hours
12.6 hours ÷ 6 hours/day = 2.1 days

Rate

Maria Pascucci, a sportswriter for the Vernon Voice, was assigned to cover the Wind River Range Hike. Pascucci figured it would take participants two days to hike the 14.5 mile route if the weather was good. If the hikers spent eight hours a day on the trail how fast did the average participant hike?

Rate = 14.5 miles ÷ (2 x 8 hr/day) = 14.5 ÷ 16
= .9 of a mile per hour (almost 1 mph)

Distance

Beulah Macklin, a reporter for the Fairhaven Falcon, was sent to the state capital to cover a legisla-

---

Common speed conversions

	Miles/hr	Meter/sec	Feet/sec	Knots
Miles/hr	—	.447	1.47	0.868
Meter/sec	0.0224	—	0.0328	0.0194
Feet/sec	0.682	.305	—	0.592
Knots	1.15	.515	1.69	—

Multiply the unit of measurement in the left column by the number in the column of the desired measurement.

tive battle over the budget. If she drove 65 mph on the interstate for five hours, how far did she travel?

Distance = 65 miles/hour x 5 hours = 325 miles

9.2 Speed, velocity, acceleration, g-force and momentum

Speed

Speed and velocity are not the same measurement, although they are often used interchangeably. Speed simply measures how fast something is going, while velocity also indicates its direction. As a reporter, it's likely that you'll only ever be interested in calculating speed.

Speed and acceleration

The speedometer on your car gives you a snapshot of the car's motion — it tells you the speed at exactly that moment. This is called "instantaneous speed." A more useful figure for a reporter is average speed, which is calculated by dividing the distance traveled by the time it took to get there.

How fast is that?

Speed of light (in a vacuum) .... 186,000 miles/sec.

Speed of sound (0° C in dry air)......... 1,088 feet/sec.

Fastest measured jet stream .... 597 feet/sec.

Earth's rotation at the equator .... 1,532 feet/sec.

Avg. speed of Moon around Earth ...... 0.634 miles/sec.

Average speed of Earth around Sun .. 18.52 miles/sec.

Formula

Average speed is another word for rate, so simply plug speed into the rate formula:

Average speed = distance ÷ time

Example

Joe Perkins was covering an old fashioned auto race for the Daily Times. The 1920s era cars traveled a 20-mile course. The winner made it in 30 minutes. What was his average speed?

Since automobile speed is measured in miles per hour, let's first convert 30 minutes to .5 hours. Then the calculation is:

Average speed = 20 miles ÷ .5 hours = 40 miles per hour

Formula

Acceleration = (ending velocity - starting velocity) ÷ time

Example

Mary Market likes her Corvette, and she knows that her vehicle can get from zero to 60 miles per hour

---

in 30 seconds. What is her rate of acceleration?

Acceleration = (60 mph - 0 mph) ÷ 30 seconds = 2 miles per hour per second

In other words, Mary's car increases in speed 2 mph every second she has the pedal floored.

The formula for acceleration can be altered to find any of its elements. If you know the acceleration and the time, you can calculate the ending speed.

Formula

Ending velocity = (Acceleration x time) + starting velocity

G-force

A g-force is an acceleration measure. One "g" represents the normal force of gravity on the Earth's surface.

Acceleration produced by gravity is measured at 9.8 meters per second per second, or 32.2 feet per second per second. Two "gs" represents 2 x 32.2 feet per second per second or 64.4 feet per second and so forth.

Typical g-forces

Protons in Fermilab accelerator	9 trillion g
Ultracentrifuge	300,000 g
Baseball struck by bat	3,000 g
Soccer ball struck by foot	300 g
Parachutist during opening of chute	33 g
Gravity on surface of Sun	27 g
Explosive seat ejection from aircraft	15 g
F16 aircraft pulling out of dive	8 g
Loss of consciousness	7 g
Gravity on surface of Earth	1 g
Gravity on surface of Moon	0.17 g

O'Hanian, Physics, 1989

Example

Jamaal Washington is writing about a new locomotive. Its stated acceleration at full throttle is .25 miles per hour per second. If the train is already moving 25 mph, and accelerates at full throttle for 2 minutes, how fast will it be going?

First we need to convert the time (2 minutes) to seconds, so our units of measurement agree.

2 minutes x 60 seconds/minute = 120 seconds

Ending velocity = (.25 mph/s x 120 seconds) + 25 mph = 55 mph

Acceleration problems can easily become complicated. What if the vehicle is accelerating uphill? What if it doesn't accelerate at a steady rate? These questions are beyond the scope of this book, but we'll look at one more acceleration problem: The speed of falling objects.

The acceleration of gravity is 9.8 meters per second per second (see the G-force sidebar). If something is falling freely in a vacuum (which means there is no friction) it will fall at that rate regardless of its weight.

Weight vs. Mass

Mass and weight are often used interchangeably, but they're not the same thing. Mass is a measure of amount. Weight is a measure of the force of gravity pulling on an object. Mass stays the same regardless of gravity. A 3 Kg bag of flour has a mass of 3 Kg on Earth and on Mars. But a 3 lb. bag of flour weighs only 1 lb. on Mars, because the gravitational pull on Mars is 1/3 of what it is on Earth.

The kilogram is the metric measure of mass, and the Newton is the metric measure of weight. A Newton is equal to 0.225 lb. on Earth.

It's simple to plug 9.8 meters per second into the acceleration formula to come up with ending velocity if you know how long something has been falling (let's assume there is no friction). But what if you only know the distance something fell? To determine the speed of that object when it hit the ground, you need to manipulate the equation for acceleration. Since you know the rate of acceleration and the distance traveled, you can figure out the ending speed. The formula would be:

Formula

Ending speed = $\sqrt{2(\text{acceleration} \times \text{distance})}$

Example

Reporter Frank Jones is writing about a high-wire acrobat who fell from 10 meters in the air into a net. At what speed was she traveling when she hit the net?

Ending speed = $\sqrt{2(9.8 \text{ m/s/s} \times 10 \text{ m})}$ = 14 meters per second

Momentum

Momentum is the force necessary to stop an object from moving. All moving objects have momentum. Momentum is the product of mass and velocity.

Formula

Momentum = mass × velocity

Example

Lori Hu covers auto racing for the Stone Sentinel. At a race last weekend a light racer weighing 132 kilograms crashed into the protective wall. What was the momentum of the vehicle as it hit the wall if it was

How loud?

The decibel is a unit of measure for the intensity of sound. One decibel is one-tenth of a Bel, a unit of measure named after Alexander Graham Bell.

The softest sound detectable by the human ear is 1 dB. Your ears hurt at 120 dB.

Decibels are a "log" scale. This means the sound intensity doubles for every 3 dB increase. For example, a sound that is 50 dB has twice the intensity as a sound that is 47 dB, and a sound that is 53 dB has twice the intensity of something that is 50 dB. That means that a 53 dB sound is four times as intense as a 47 dB sound (because the intensity has doubled twice).

However, sound intensity is not directly equivalent to "loudness." "Loudness" is a subjective measure of how a person's ears perceive sound. An accepted rule of thumb is that to double the loudness of something the intensity has to be multiplied by 10. Think of it this way: Two trumpet players will produce twice the sound intensity of one trumpet player, but your ears will not perceive a doubling in loudness (if your ears did, you couldn't stand to listen to a full orchestra!). It would take 10 trumpet players for the loudness to be roughly double one trumpet player.

A 10 dB increase is a 10 times increase in intensity; so a 10 dB increase means a sound has roughly doubled in loudness.

Let's say you get a press release stating that a new piece of equipment at a factory will increase the sound intensity on the shop floor from 50 dB (about the sound of a normal conversation) to 60 dB. Your readers might perceive that as a rather small increase, but in reality it would be doubling the noise on the shop floor.

Imagine a conversation with two people talking (about 50 dB); now imagine there are 20 people all talking at once (about 60 dB). That's how much louder the new factory will equipment will be.

traveling 105 mph?

First let's convert miles per hour to kilometers per hour, so all our units of measure are metric.

105 mph x 1.6 = 168 kph

132 kilograms x 168 kilometers per hour = 22,176 kilogram kilometers per hour

Skill Drills

1. Mary Brewer, a reporter for the Presque Parcel, was assigned to write a feature story about life on a cruise ship catering to college students on Spring Break. The ship was located 112 miles off the coast. How many nautical miles from shore was the ship? While on the ship she learned that a supply boat traveled the same distance at 15 knots. How fast did the supply boat travel (expressed in miles per hour)?

2. Tilman Vookles, a sportswriter with the Alpha Anchor, decided to write a first person account of participating in the Chicago marathon. The marathon covers 26.2 miles. If it took Tilman 4 hours and 34 minutes to complete the race, what was his speed?

3. Ruben Richardson, a feature writer with the Grant Guardian, is given the opportunity to ride in a racecar at the local racetrack. The car has a mass of 785 kilograms and is traveling at a velocity of 48 kilometers/hour. What is its momentum?

4. Auto reporter Jim Fletcher wants to write about the insurance industry crash tests. The press release about a particular test states that a truck was crashed into a wall at 50 mph and suffered minimal damage. Fletcher investigates the numbers behind the test and learns that the truck was put on a track that was 1/10th of a mile long and accelerated at a constant acceleration of 1.5 miles per hour per second. Was the truck really traveling 50 mph when it hit the wall?

---

5. Gene Kimball, the police reporter for the Talbot Times, covered a stolen car chase that resulted in a collision on Camilla Highway. The state police clocked the car going at 81 mph 13 minutes into the chase. Two minutes later the car was clocked at 123 mph. Assuming the car accelerated at a constant rate, what was its acceleration rate for those two minutes?

6. A camera man dropped his 20-kilogram camera out of a traffic helicopter during a report on a traffic jam. If the helicopter was hovering 50 meters above the road, how fast was the camera moving when it hit the ground? What was its momentum?

Learning Challenge

Visit a local amusement park and obtain data on roller coaster rides. What is the total length of the track? Using a stopwatch, time how long it takes a car to travel the entire rollercoaster. Knowing the distance traveled and the time, calculate the average speed.

Now ask how high the highest point of the roller coaster is, the length of track from that point to the bottom of that hill, and the maximum speed of the coaster. Assuming the coaster stops momentarily at the top of its biggest hill and reaches its maximum speed at the bottom, calculate the acceleration going down the hill. Convert that to "gs."

Chapter 10

Area Measurements

Measurement statistics pop up in all kinds of news stories. Knowing how to express measurements in an accurate and clear way is vital to good journalism.

There are two ways to explain measurements, and each has its place.

One way is by analogy. "The casino is the size of a football field," "the bridal magazine was as thick as a phonebook," "the tree was as tall as a four-story building."

Analogies are great for illustrating measurements that may be otherwise meaningless. For example, writing that a tree is as a tall as a four-story building is probably more effective than saying a tree is 40 feet tall. Readers can visualize a

Numbers in the news...

Looming Large

Larry Fulghum can weave quite a story about his work-place.

The new $8.5 million loom he operates at Weavexx is the largest needling instru-ment of its kind in the Western Hemisphere.

The loom converts batt rolls of nylon — produced else-where in the 200,000-square-foot plant — into a myriad of felt textures. The most com-mon run from 80 to 120 feet.

Weavexx expanded its plant by 7,500 square feet to incorporate the loom.

Daily Journal,Tupelo, Miss. Nov. 20, 2002

four-story building with a tree beside it, and they get an idea of how tall the tree is. Few people can readily visualize something that is 40 feet tall by itself.

But analogies only work if the readers understand the comparison. For example, if you are writing for a national trade magazine, don't compare the size of a new industrial park to the municipal park in your hometown. Only the readers who live there will understand the analogy.

Analogies also fail when exact measurements are essential. If you want to convey the size of a new parking lot going up in a residential neighborhood, the surrounding home owners probably want to know exactly how big the lot is. Don't say it's "about the size of a football field."

In such cases it's best to use the second method to explain measurements: with simple, accurate numbers.

This chapter will explain some basic measurements that reporters should be able to calculate.

It's good to know how to calculate perimeter when you're writing articles about new developments, construction projects and other topics involving the size of things on the ground.

Area is key to stories about anything that deals with the size of surfaces. Real estate and construction stories obviously fall into this category, but so do countless technical stories, feature articles, and sports reports.

Knowing how to accurately calculate square feet and square yards pays off when you're checking official reports of size, measuring areas on your own, or otherwise dealing with these figures.

Finally, understanding circumference and radius is important when working on any story involving circular areas.

10.1 Perimeter

Formula

If the object is a square or a rectangle use the formula:

Perimeter = (2 x length) + (2 x width)

Example

The paved portion of the courtyard outside the provost's office is a rectangle with a length of 9 feet and a width of 5 feet. What is the perimeter?

Perimeter = (2 x 9) + (2 x 5) = 28 feet

If the object has an irregular shape, add the lengths of all sides to determine the size of the perimeter.

Example

The parks department has created a new flower bed beside village hall that has two sides measuring 7 feet, one side measuring 9 feet and one side measuring 6 feet. What is the perimeter of the flower bed?

7 + 7 + 9 + 6 = 29 feet

10.2 Area of squares and rectangles

Formula

The formula for the area of a rectangle or a square is the same:

Area = length x width

Example

After the Winkler Town Hall was destroyed by fire, city officials negotiated with several developers to rent temporary office space. Ardis Inc. offered to rent the town space in its Brenton Place Tower for $12 per square foot, per year. Each office is 40 by 90 feet. How much would one office cost the city per year?

40 x 90 = 3,600 square feet
3,600 x $12 = $43,200

Formula

To find the area of a triangle:

Area = .5 base x height

(Use the two shortest sides as base and height)

Example

Bern Community Church wants to buy a triangular tract of land from the city to have additional parking for Sunday mornings. The section adjoining the church's existing parking lot is 118 feet. The length (height) of the tract along the sidewalk is 250 feet. What is the area of the tract?

.5 x 118 x 250 = 14,750 square feet

10.3 Square feet, square yards

Small spaces are measured in square inches or square feet. Larger areas, such as parking lots, are measured in square feet, square yards, or square rods. Fields and farms are measured in acres. Cities, states, and counties are measured in square miles.

144 square inches = 1 square foot
9 square feet = 1 square yard
30 square yards = 1 square rod

---

160 square rods   = 1 acre
1 acre = 43,560 square feet
640 acres = 1 square mile

Example
How many square rods in 9 acres?
9 acres x 160 square rods/acre = 1,440 square rods

10.4 Radius, circumference, area of circle

The radius of a circle is the distance from any edge to the middle. Knowing the radius is key to finding the circumference (the distance around it) or area of a circle. Another important figure in calculations about circles is pi (π), which is rounded off to 3.14.

Formula
Circumference = 2π x radius

Example
Reporter John Freeman wants to know the distance around a round flower patch in Hammond Park. The distance from the edge of the patch to the middle is 10 feet.

Circumference = 2π x 10 = 6.28 x 10 = 62.8 feet

Formula
Area = π x radius^2

Example
Now Freeman wants to know the area of the flower patch.

Area = π x 10^2 = 3.14 x 100 = 314 square feet

Skill Drills

1. What is the perimeter of a football field 100 yards long with two end zones of 10 yards each and a width of 50 yards? What is the area of that field?

2. For his Introduction to Design class, Li-Yong Kao had to design a usable object with a perimeter of 56 inches. If his design was a square, what was the size of each side?

3. Nancy Mellon, head of the state conservation department, announced plans for a new forest preserve in Garner County. The site, beside the Buchanan River, measures 31 miles along the river and extends 11 miles due west. A new road will surround the site on the three sides not bordered by the river. Assuming the river runs straight north to south, how long will the road be?

4. Maudine Wallace wants to donate land to the university for a wildlife retreat near the Chickasaw River. She learned that a tract of land along Old Mill Road is available. The land extends back from the road for six miles and along the road for 1.2 miles. How many square feet of land are available?

5. The Alzheimer Center on Bynum Road has prepared a new triangular flower bed near the rear patio. One side of the flower bed is the same width as the patio, 23 feet, and another side runs along an adjacent sidewalk for 42 feet. What is the area of the flower bed?

6. After building a new highway, the Illinois Department of Transportation found that it had a surplus triangular lot adjacent to the highway. Land in the neighborhood sells for $20,000 per acre. If the lot has a base width of 500 feet and a length of 1,250 feet, how much can IDOT expect to earn from sale of the land?

Learning Challenge

Each year colleges renovate buildings, government expands offices, schools clamor for more space. Public officials release figures attesting to the cost of improvements per square foot. But does anyone ever check?

1. Calculate the actual square footage of a recent development, such as a playground or parking lot. Do the published measurements match the class measurements? What might account for the differences, if any?

2. Calculate the cost per square foot for renovations to specific buildings. Compare the results to published reports. Do the published costs per square foot match the class measurements? What might account for the differences, if any?

3. Create a clip file of newspaper stories using square footage, and a file of stories where doing the math would have resulted in a more accurate, better-written story.

Chapter 11

Volume Measurements

Volume measurements play a key role in many articles. How many tons of rock salt does a town need to handle a rough winter? How much salt is needed per mile of road? How much salt can each truck hold? All sorts of details like these can add important context to articles.

In the business world terms like ton, box, barrel and cord take on specific meanings. Goods often are sold based on volume, and since volume measurement terms can vary based on the market, knowing the measurement device is as important as the selling price.

This section looks at various volume measurements — from the size of a cup of milk to measurements used to specify the volume of building materials.

11.1 Liquid measurements

Liquid measurements apply to liquids in recipes, bodies of water, and other fluids.

Common liquid conversions

2 tablespoons = 1 fluid ounce

1/2 pint = 8 ounces, or one cup

1 pint = 16 ounces, or two cups

2 pints (32 ounces) = 1 quart

2 quarts (64 ounces) = 1/2 gallon

4 quarts (128 ounces) = 1 gallon

1 U.S. standard barrel = 31.5 gallons

1 U.S. gallon = 4/5 Imperial gallon

British or Canadian barrel = 36 Imperial gallons

(When used to reference crude oil on the international market, a barrel contains 42 U.S. gallons or 35 Imperial gallons)

Example

How many servings are there in a gallon of milk if each serving is 8 ounces?

Reduce the amounts to the same measurement form. Multiply to move up the scale, divide to move down the scale.

1 gallon = 4 quarts = 128 ounces
128 ounces ÷ 8 ounces (one serving) = 16 servings

Example

Longleaf Bottled Water Co. announced that it would expand its international market next year by shipping more cans of water to Asia. The company packaged 1.26 billion 12-ounce cans at its plant in Upper Black Eddy last year. If each case contains 24 cans, and one barrel equals 13.77 cases, how many barrels of water were used at the plant?

The solution calls for reducing the problem to similar math forms and using multiplication and division to get the answer.

24 cans/case x 12 oz = 288 oz/case

288 oz/case x 13.77 cases/barrel = 3,965.76 oz/ barrel

1.26 billion x 12 ounces = 15.12 billion ounces were packaged

15.12 billion ounces packaged ÷ 3,965.76 oz/barrel = 3.81 million barrels

---

Lights, camera...electricity bill

How much electricity did that film crew consume when it was filming the ad on campus? Even if you don't get the electricity bill, you can still get a rough idea. First, find the maximum wattage listed on the equipment they used. Then estimate the amount of time in hours the equipment was on. Plug those numbers into this formula:

Wattage x time = Energy consumed, in watt-hours.

Your local utility probably measures electricity in kilowatt hours, so convert your answer to KWH by dividing by 1,000. Now call the utility to get the cost of electricity per KWH. Multiply times the KWH, and you know roughly how much higher the college's electricity bill will be this month because of the filming.

11.2 Rectangular solid

Formula

Volume = length x width x height

Example

What is the volume of a shoebox 11.5 inches long, six inches wide and four inches tall?

11.5 in. x 6 in x 4 in. = 276 cubic inches

11.3 Cord

Firewood is sold by a measurement called a cord. A cord is defined as 128 cubic feet when the wood is neatly stacked in a line or row. A standard cord would be a stack of wood 8 feet long, 4 feet wide and 4 feet high.

11.4 Ton

There are three different types of tons:
■ A short ton is equal to 2,000 pounds
■ A long ton, also known as a British ton, is equal to 2,240 pounds
■ A metric ton is equal to 1,000 kilograms, or 2,204.62 pounds

Ton conversions

Short to long ton: Multiply by .89

Short to metric ton: Multiply by .9

Long to short ton: Multiply by 1.12

Long to metric ton: Multiply by 1.02

Metric to short ton: Multiply by 1.1

Metric to long ton: Multiply by .98

Skill drills

1. Brooks Libbey, a reporter for the Johnson Journal, was investigating federal kindergarten milk payments to the Reaves School. She decided to calculate how many gallons of milk were needed each week. In looking at the attendance records she found that the school averaged 302 children present each day Monday through Friday. How much milk did the school need each week if every child was present and was served an 8-ounce glass of milk?

2. Ivan Gorshkov, the energy writer for the Copeland Courier, was given an assignment to find the best deal on 1,000 barrels of oil. Ivan found he could buy oil on the Internet. Which was the best deal?
A. 59 cents a gallon from a source in the United States
B. 39 cents a gallon from a source in Canada
C. $1.19 a gallon from a source in Saudi Arabia

3. Warren Korp, a consumer reporter for the Stillman Signal, decided to write a story about the growth of highway rental storage facilities. At the Asmore Storage Facility the most popular rental was a storage unit that measured 6.5 feet by 10 feet by 8 feet. How many cubic feet was the storage area?

4. Commercial fishing reporter Hal Austell accompanied the crew of the Ann Henry on a fishing trip off the Grand Banks for a feature story. The crew caught 22,000 pounds of fish on the trip. How many short tons of fish did he catch? Convert the number to long tons.

Learning Challenge

1. How much water drips from a leaking faucet in one day? In one year? Put a measuring cup under a dripping faucet and collect water for 15 minutes. If each of the estimated 90 million households in the United States had one faucet dripping at this rate, how much water would be wasted in a year?

2. Price the cost of firewood at various locations around town. Measure the stacks yourself to determine the accuracy of the retailer's volume measurements.

3. Find out how many tons of goods are shipped from your local port, airport or tractor-trailer depot. How have the numbers climbed or fallen since five years ago? What factors are behind the shift? Do your local shippers measure quantities in short tons, long tons or metric tons? Why do they use the method they use?

Chapter 12

The Metric System

Do you think the metric system is hard to understand? You have good company among Americans, but virtually the rest of the world uses the metric system for nearly every type of measurement. The metric system is an important part of international commerce and science, and as a journalist you should have a good grasp of it.

The international decimal-based metric system is based on multiples of 10. Every measurement uses standard language for each level (giga, mega, milli, micro, etc.). Thus, once one learns the language, it becomes simple to use the metric system. In fact, it is much more logical than the "English" system of weights and measures Americans now use.

Despite Americans' stated dislike of the metric system, we already use the metric system on a regular

Numbers in the news...

Save us From Metric

The Treasury Department's Bureau of Alcohol, Tobacco and Firearms has saved us from centiliters. And maybe even dekameters, decimeters and hectometers. In response to a petition from a vintner in Old Brookville, N.Y., to label 750-milliliter, or "750 ml," wine bottles as "75 cl," ATF said, charitably, that this would just be too confusing for American consumers....

Washington Post, July 9, 1999

basis — in sports, for soda sold in liter bottles and in medicines, among other items. Thus, familiarity is growing, if not acceptance.

12.1 Definitions

The meter is the basic unit for length. It is based on a measure that equals one 10-millionth of the distance from the North Pole to the equator along the meridian running near Dunkirk in France and Barcelona in Spain.

Mass is also derived from the meter. One gram is the mass of one cubic centimeter of water at 0 degrees Celsius. The International Bureau of Weights and Measures keeps a cylinder of platinum-iridium alloy that has a mass of exactly one kilogram (1,000 grams) as the official standard.

The metric unit of force is the Newton. One Newton is a force that, applied to a one-kilogram object, will give the object an acceleration of one meter per second per second. Because the force of gravity is what gives objects weight, the Newton also can be used as a measure of weight. One kilogram weights 9.8 Newtons on Earth.

Volume traces its origins to the cubic decimeter, a cube 0.1 meter on each side.

12.2 Basics

Because the metric system is based on the decimal system, users can change from one unit to another by multiplying or dividing by 10, 100, 1,000 or other multiples of 10. Each unit is 10 times as large as the next smaller unit.

The unit names are meter (length), gram (mass) and liter (volume).

Prefixes, when added to a unit name, create larger or smaller factors. The prefixes and their numerical values are:

micro (1 millionth)   0.000001
milli (1 thousandth)   0.001

Measurement comparisons

Name	Symbol	Approximate size
Length		
meter	m	39.5 inches
kilometer	km	0.6 mile
centimeter	cm	0.4 inch
millimeter	mm	thickness of a dime
Area		
hectare	ha	2.5 acres
square meter	m^2	1.2 square yards
Mass (right column is comparable weight on Earth)		
gram	g	weight of a paper clip
kilogram	kg	about 2.2 pounds
metric ton	t	long ton (2,240 lbs.)
Volume		
liter	L	1.2 ounces
milliliter	mL	1/5 teaspoon

Temperature	Celsius (C)	Fahrenheit (F)
water freezes	0	32
water boils	100	212
normal body temp.	37	98.6
comfortable room temp.	20-25	68-77

centi (1 hundredth)   0.01
deci (1 tenth)   0.1
no prefix   1.0
deka   10
hecto   100
kilo   1,000
mega   1,000,000
giga   1,000,000,000
tera   1,000,000,000,000

For example, using the root measure meter, a millimeter is 1/1000 of a meter, a centimeter is 1/100 of a meter and a kilometer is 1000 meters. Turn the figures around and 1,000 millimeters = 1 meter; 100 centimeters = 1 meter; and 1,000 meters = 1 kilometer.

To change a unit to a larger unit divide the given unit by a desired multiple of 10.

Example
670 meters ÷ 1,000 = .67 kilometers

To change a unit to a smaller unit multiply the given unit by the desired multiple of 10. Users can also move the decimal point one place to the right for each zero in the multiplier, or if the number lacks a decimal point, add one zero to the number for each zero in the multiplier.

Example
4.80 meters x 1,000 = 4,800 millimeters

12.3 Length

Formula
To convert American lengths to metric:

Multiply

— inches by 25.4 to find millimeters or 2.5 to find centimeters

— feet by 30 to find centimeters or 0.3 to find meters

— yards by 90 to find centimeters or 0.9 to find meters

— miles by 1.6 to find kilometers

Example
Angie Calhoun, a community reporter for the Rowland Register, was reading the building specifications for an indoor swimming pool proposed for the Youth Center. The specifications called for a pool 80 feet long. Because she knew the high school wanted to use the pool for a swim team, she converted the feet into meters to determine if the length of the pool met

competition requirements calling for a length of 25 meters. Did the pool specifications meet the requirements?

80 feet x 0.3 = 24 meters

Formula

To convert metric lengths to American:

Multiply

— millimeters by 0.04 to get inches
— centimeters by 0.4 to get inches
— centimeters by 0.033 to get feet
— meters by 39 to get inches
— meters by 3.3 to get feet
— meters by 1.1 to get yards
— kilometers by 0.62 to get miles

Example

The Meadowbrook Running Club announced that it would sponsor a 5 km run to raise funds for the town's Performing Arts Center. Zach Ragland, a feature writer for the Meadowbrook Monitor, converted the kilometers to miles.

5 kilometers x 0.62 = 3.1 miles

Sports distance conversions (approximate)

100 yards = 91 meters
220 yards (1/8 mile) = 200 meters
440 yards (1/4 mile) = 400 meters
880 yards (1/2 mile) = 800 meters
1 mile = 1,600 meters
1 furlong = 220 yards = 1/8 mile = 200 meters

12.4 Area

Formula

To convert American area measurements to metric:

Multiply

— square inches by 6.5 to find square centimeters

— square feet by 0.09 to find square meters

— square yards by 0.8 to find square meters

— square miles by 2.6 to find square kilometers

— acres by 0.4 to find hectares

Example

Elliott Wolfe, a suburban reporter for the Miller Merchandiser, was working on a story about planned unit developments. He noted that in the last 10 years three PUDs were built in and around Miller. The largest of the developments measured 1,700 acres. He converted the size to hectares.

1,700 acres x 0.4 = 680 hectares

Formula

To convert metric measurements to American:

Multiply

— square centimeters by 0.16 to get square inches

— square meters by 11 to get square feet

— square meters by 1.2 to get square yards

— hectares by 2.5 to get acres

— square kilometers by 0.4 to get square miles

Example

Elisha Hollandale, a reporter for the Hanover Herald, spent the day with volunteers building a home for a low-income family. The specifications for the house were given in metric. She converted them to the American system. If the house measured 113 square meters, what was its size using the American system?

113 square meters x 11 = 1,243 square feet

12.5 Mass

Formula

To convert American mass measurements to metric:

Multiply

— ounces by 28 to find grams

— pounds by 0.45 to find kilograms

— pounds by .07 to get stones

— short tons (2,000 pounds) by 0.9 to find metric tons

Example

Stan Mosaic, a police reporter for the Edison Escort, read in the daily crime report that the Edison police had arrested a man on the Green Parkway with 29 pounds of marijuana in the car. Convert the figure to kilograms.

29 lb x 0.45 = 13.05 kg

Formula

To convert metric mass measurements to American:

Multiply

— grams by 0.035 to get ounces

— grams by 0.002 to get pounds

— kilograms by 35 to get ounces

— kilograms by 2.2 to get pounds

— metric tons by 1.1 to get tons

Example

Lily Ray, a general assignment reporter for the Kent Kernel, was assigned to report on a traffic accident caused when a truck hauling 3 metric tons of frozen food skidded through a traffic light and overturned in a ditch. How many tons of goods were lost using the American measurement system?

3 metric tons x 1.1 = 3.3 tons

12.6 Volume

Formula

To convert American volume measurements:

Multiply

— teaspoons by 5 to find milliliters

— tablespoons by 15 to find milliliters

— fluid ounces by 30 to find milliliters

— cups by 0.24 to find liters

— pints by 0.47 to find liters

— quarts by 0.95 to find liters

— gallons by 3.8 to find liters

— cubic feet by 0.03 to find cubic meters

— cubic yards by 0.76 to find cubic meters

Formula

To convert metric volumes to American forms:

Multiply

— milliliters by 0.034 to get fluid ounces

— milliliters by 0.002 to get pints

— liters by 2.1 to get pints

— liters by 1.06 to get quarts

— liters by 0.26 to get gallons

— cubic centimeters by 0.06 to get cubic inches

— cubic meters by 35 to get cubic feet

— cubic meters by 1.3 to get cubic yards

Example

Jared Squires, a parks and recreation reporter for the Anderson Announcer, noted that the park commission put down 120 cubic feet of pea gravel under the new playground in Babin Park. How many cubic meters did the park commission use?

120 cubic feet x 0.03 = 3.6 cubic meters

12.7 Temperature

Formula

To convert Fahrenheit to Celsius:

Celsius = .56 x (Fahrenheit - 32)

To convert Celsius to Fahrenheit:

Fahrenheit = (1.8 x Celsius) + 32

12.8 Style rules

The National Institute of Standards and Technology, a division of the U.S. Department of Commerce, suggests the news media use the following style rules.

Capitals

■ Units: The names of all units start with a lower case letter except, of course, at the beginning of the sentence. There is one exception: in "degree Celsius" (symbol °C) the unit "degree" is lower case but the modifier "Celsius" is capitalized.

■ Symbols: Unit symbols are written in lower case letters except for liter and those units derived from the name of a person or country, such as Newton (N).

■ Prefixes: Symbols of prefixes that mean a million or more are capitalized and those less than a million are lower case: M for mega (millions), m for milli (thousandths).

Plurals

■ Units: Names of units are plural only when the numerical value that precedes them is more than 1. For example, 0.25 liter or 1/4 liter, but 250 milliliters. Zero degrees Celsius is an exception to this rule.

■ Symbols: Symbols for units are never pluralized (250 mm = 250 millimeters).

Spacing

A space is used between the number and the symbol to which it refers. For example: 7 m, 31.4 kg, 37 °C.

When a metric value is used as a one-thought modifier before a noun, hyphenating the quantity is not necessary. However, if a hyphen is used, write out the name of the metric quantity with the hyphen between the numeral and the quantity. For example:

a 2-liter bottle, or a 2 L bottle (not a 2-L bottle);

a 100-meter relay, 100 m relay;
35-millimeter film, 35 mm film.

In names or symbols for units having prefixes, there is no space between letters making up the symbol or name. Examples: milligram, mg; kilometer, km.

Period
Do not use a period with metric unit names and symbols except at the end of a sentence.

Decimal point
The dot or period is used as the decimal point within numbers. In numbers less than one, zero should be written before the decimal point. Examples: 7.038 g; 0.038 g.

Skill Drills

1. Sara Wilson, a sophomore journalism major, was applying to spend her junior year in England on an exchange program. The application asked her to list her height in metric. She is 5'5" tall. How tall is she on the metric scale? Hint: Convert her height to inches first.

2. Connie Florendo, a police reporter for the Phifer Flute, was assigned to cover the story of an abandoned car on the interstate that police discovered housed a bomb and multiple high-powered weapons. The driver of the car abandoned the vehicle after he hit the stop sign and slid into a ditch. Police officers on the scene told Florendo that when they exploded the bomb debris scattered 2.6 meters. How far did the debris scatter in feet?

3. Archibald Gill, an investigative reporter for the Fowler Flag, noted in city records that the city purchased a 50-gallon fuel tank for the maintenance yard. Records indicated that the tank was filled twice a week. However, the report from the alderman in charge of the maintenance yard said that the new tank held 150 liters and was being filled three times a week. How much fuel was being consumed in gallons based on Gill's research? How much fuel was being consumed in liters based on the alderman's report? What is the difference?

4. Janet Moore, an investigative reporter for the Perskie Peripheral, was looking over records in the planning office and noticed that a nationally-known discount retail store had purchased 4,000 square yards near the

interstate to build a new store. How many square meters were purchased?

5. Scott McDonald, a feature writer for the Hughes Herald, was assigned to do a story on can recycling. He spent part of the day at the Joyner Recycling Plant. While he was there he watched 15 short tons of cans being crushed and molded into shipping containers. How many metric tons were crushed and molded?

6. Tia Cooper, the food writer for the Kenton Key, spent a week translating metric recipes from her immigrant grandmother's cookbook into the American system. A 2.25 kg bag of flour weighs what in the American system?

7. Spencer Emmett, a reporter for the Davis Drum, was working on a story about a heat wave in Davis. The temperature hit above 100 degrees 14 days in a row. He found that the world's highest temperature occurred in el-Azizia, Libya. The temperature was reported as 58 degrees Celsius. What was the temperature in Fahrenheit?

Learning Challenge

The metric system was created at the end of the 18th century to provide a consistent system of units for measurement. Previously, each area had its own units inherited from earlier times.

The Metric Program of the U.S. Department of Commerce's National Institute of Standards and Technology coordinates the metric transition activities of federal agencies and works with states and regional authorities to encourage the use of the metric system. The program estimates that more than 30 percent of American products are metric or labeled in metric units.

Try to have a "metric-only" day. When you shop, buy and compare items using their metric measurements. If you use a recipe, convert the measures to metric. If you drive somewhere, calculate your fuel consumption in kilometers per liter rather than miles per gallon. When you read the thermometer, only look at the Celsius numbers. At the end of the day, write an article about the experience.

Appendix 1

Math Basics

Don't be embarrassed if you've forgotten the basics of math. You probably learned most of them in elementary school, and that was a long time ago! On the following pages is a primer on some basic math concepts that will help you as you work through the rest of this book.

A1.1 Equality, addition, subtraction

The equal sign, =, is used to indicate that two items are the same. That two items are to be added is denoted by the plus sign, +, and subtracted by the minus sign, -.

A1.2 Multiplication

There are several ways to indicate that something is to be multiplied. The most common, and the way we mostly use in this book, is with an "x" symbol.

$$5 \times 2 = 10$$

If the formula calls for a number to be multiplied by some numbers or figures inside parentheses, then you usually just put the number up against the parentheses.

$$5(2a + 3b) = 10a + 15b$$

Notice that the 5 was multiplied by each of the figures inside the parentheses, not just the first one.

Speaking of parentheses, do the work inside the parentheses before the rest of the equation.

$$(4 + 6) \div (3 - 1) = 10 \div 2 = 5$$

If a formula calls for you to add or subtract some figures and multiply or divide others, do the multiplication or division first (unless something is in parentheses; then do that first).

$$5 + 6 \times 3 = 5 + 18 = 23$$

A1.3 Division

Division can be indicated in two basic ways. The most common is with the division sign, $\div$.

$10 \div 2 = 5$

Another common way is to put the number to be divided on top of the other number, or to put a slash between them.

$10/2 = 5$ $\qquad \dfrac{10a - 6b}{2} = 5a - 3b$

Notice in the second equation, the 2 divided both numerators above it.

A1.4 Squares, square roots, to the power of

To square something, multiply it by itself. The square of 2 is:

$2 \times 2 = 4.$

You can indicate a figure is to be squared by floating a little 2 next to it. For example, 5 squared is written 5^2.

$5^2 = 5 \times 5 = 25$

The square root of something is the number that, when multiplied by itself, equals the original number. The square root of 16 is 4, because $4 \times 4 = 16$. In formulas, square root is symbolized by this: $\sqrt{}$

$\sqrt{16} = 4$

"To the power of..." is another way of indicating something needs to be multiplied by itself. For example, "to the second power" is the same as being squared. "To the fourth power" means the figure should be multiplied by itself four times. For example,

5 to the fourth power is:

$5^4 = 5 \times 5 \times 5 \times 5 = 625.$

A1.5 Fractions

Fractions consist of two numbers: the numerator on top and the denominator below. In the fraction 2/5, the 2 is the numerator and 5 is the denominator. As we discussed in the section on division, that line separating the numbers means "divided by." Thus, this same fraction could be expressed as 2 divided by 5, which equals 0.4.

Some day you may be asked to add or subtract fractions. Let's say a landowner says 2/5 of a certain parcel of land belongs to him, but his neighbor claims 4/5 of that parcel. Since the denominator of both fractions is 5, it's easy to add them:

2/5 + 4/5 = 6/5

Six-fifths is more than one (because 6 divided by 5 equals 1.2), so at least one of the landowners is wrong.

But what if you have a fraction with unequal denominators? Then you need to make the denominators equal before you add the fractions. Let's say Eckford Media Corp. claims 2/3 of the marketshare for newsprint, and Zigma Co. says it holds 1/5 of the marketshare for the same product. Who holds the larger marketshare, and by how much?

The quickest way to make the denominators equal is to multiply them by each other. Then you need to multiply the numerators by the denominator of the other fraction.

2/3 = Eckford Media Corp. marketshare
1/5 = Zigma Co. marketshare

Common denominator = 3 x 5 = 15

Now multiply the numerator 2 in the Eckford Media Corp. fraction by 5 (the old denominator of the Zigma Co. fraction), and put that over the new denominator: 10/15. Do the same with the Zigma Co. fraction, and you end up with 3/15.

Now you can easily compare the figures: Eckford Media Corp. holds 10/15 of the market, and Zigma Co. holds 3/15. How much of the market do the companies control together? Just add the figures:

10/15 + 3/15 = 13/15.

A1.6 Rounding

In a news report, rarely will you need to write numbers with more than two digits to the right of the decimal point. Calculations may provide more than two digits to the right of the decimal, so round off. If the last digit is 5 or higher, round up. If it's 4 or lower, round down.

Example:

4.767 rounds to 4.77

55.314 rounds to 55.31

12.255 rounds to 12.26

Skill Drills

1. Village News Reporter Brenda Russell was reading an annual report for Chicago Hot Dog Co. She came across an equation with the same figures written two different ways: 5 x 3 and 5(3). Are they equal?

2. Copy editor Steve Renkowitz is checking the math on a report issued by Hubert Chemical Corp. It says the result of certain scientific function can be expressed as 4 + 13(7 + 5). What is the solution to this equation?

3. Wealthy philanthropist Walter Purnell announces that he will make a donation each year to the Cincinnati City Museum. He will donate $1 million times the square root his daughter's age each year. She is now 25. How much will Mr. Purnell donate this year? In the year his daughter turns 36?

4. Science reporter James England finds that the formula for a high school physics project calls for taking the diameter of a wheel to the nth power, where n equals the number of wheels. One student's vehicle had three wheels, each with a diameter of 6 inches. What is 6 to the third power?

5. Shameeka Wilson and her brother Colin Wilson own a grocery store together with some friends. Shameeka owns 1/2 of the store and Colin owns 1/5. Shameeka buys Colin's share. What fraction of the business does she now own? What fraction to the other owners own?

Learning Challenge

Basic mathematics show up in newspaper and magazine articles every day. From basic addition to decimals and percentages, journalists routinely put math into their articles.

Examine a metropolitan daily newspaper from cover to cover and note every instance of basic mathematics.

Prepare a "mathematical examination" of the issue, noting which sections have the most math, any mistakes you find in the math, and any articles that you feel would have been improved with more math or better explanations of the math already there.

Appendix 2
Online Math Resources

Math tips

Bill Dedman's Power Reporting .... powerreporting.com

Internet Mathematics Library ....www.mathforum.com/library

Dr. Math ....www.forum.swarthmore.edu/dr.math/

American Mathematical Society ....www.ams.org

Niles Online ....www.robertNiles.com

Consumer credit calculators

Auto ....www.auto-loan.com

Auto, credit and mortgage .... www.ineedcredit.com

Auto, credit and mortgage .... www.anyloan.com

Cost-of-living calculators

Bureau of Labor Statistics ....www.bls.gov

Columbia Journalism Review .... www.cjr.org/resources/

Consumer Price Index .... www.westegg.com/inflation

Cost of Living .... www.newsengin.com/neFreeTools.nsf/CPIcalc

Currency calculators

Expedia .... www.expedia.com

World currency   www.rubicon.com/passport/currency/currency.html

Finance and investment calculators

Kiplinger .... www.kiplinger.com/tools/index.html

U.S. Securities and Exchange Commission .... www.sec.gov/
    investor/tools/mfcc/mfcc-int.htm

Math and statistics calculators

Basic math .... www.math.com/students/tools.html

ConvertIt .... www.convertit.com/Go/convertIt/Calculators/Math

Math Goodies .... www.mathgoodies.com/calculators

Percent change .... www.newsengin.com

Scientific Calculator ....www.scientificcalculator.com

---

Mortgage calculators

Buying a home .... www.interest.com/hugh/calc
Payment calculator ....list.realestate.yahoo.com/re/calculators/
    payment.html

Weights and measures calculators

Metric .... www.worldwidemetric.com/metcal.htm
Online Conversion ....www.onlineconversion.com

Miscellaneous calculators

Perpetual calendar .... www.world-calendar.com
Public debt .... www.publicdebt.treas.gov/opd/opdpenny.htm
Time zone calculator .... www.worldtimezone.com
U.S. Naval Observatory master clock ....www.tycho.usno.navy.mil/
    what.html

Collections of calculators

Calculator.com ....www.calculator.com
George Landau's News Engin Inc. .... www.newsengin.com

Index

Formula Quick Reference

Acceleration
Acceleration = (ending velocity - starting velocity) ÷ time

Area, circle
Area = π x radius^2

Area, square or rectangle
Area = length x width

Area, triangle
Area = .5 base x height (Use the shortest sides as base and height)

Assessed value
Assessed value = Appraisal value x rate

Assets
Assets = Liabilities + Equity

Bond cost
Bond Cost (interest) = amount x rate x years

Circumference
Circumference = 2π x radius

Current ratio
Current ratio = assets ÷ liabilities

Current yield
Current yield = coupon ÷ price

Deaths per 100,000 people
Deaths per 100,000 = (Total deaths ÷ total population) x 100,000

Debt-to-asset ratio
Debt-to-asset ratio = total debt ÷ total assets

Debt-to-equity ratio
Debt-to-equity = total liabilities ÷ total equity

Distance, rate and time

Distance = rate x time   Rate = distance ÷ time
Time = distance ÷ rate

EBITA

Net profit (EBITA) = gross margin - overhead

Gross Domestic Product (GDP)

C = consumer spending on goods and services
I = investment spending   G = government spending
NX = next exports (exports minus imports)

GDP = C + I + G + NX

Gross margin

Gross margin = Selling price - cost of goods sold

Gross profit

Gross profit = Gross margin x number of items sold

Inflation, adjusting for, from the past

A = Target year value, in dollars   B = Starting year value, in dollars
AC = Target year CPI   BC = Starting year CPI

A = (B ÷ BC) x AC

Inflation, adjusting for, into the future

C = cost after one year   K = original cost   I = inflation rate

$C = K(1 + [I \div 12])^{12}$

Inflation rate, annual

A = Annual Inflation Rate   B = Current month CPI
C = CPI from same month in previous year

A = (B - C) ÷ C x 100

Inflation rate, monthly

Monthly Inflation Rate =
(Current CPI - Prior Month CPI) ÷ Prior Month CPI x 100

Interest
Interest = principal x rate x time (in years)

Interest-bearing account, balance after one year
B = balance after one year   P = principal
R = interest rate   T = times per year the interest is compounded

$$B = P(1 + R \div T)^T$$

Mill levy
Mill levy = Taxes to be collected by the government body ÷ assessed valuation of all property in the taxing district

Momentum
Momentum = mass x velocity

Monthly payment
A = monthly payment   P = original loan amount
R = interest rate, expressed as a decimal and divided by 12
N = total number of months

$$A = [P \times (1 + R)^N \times R] \div [(1 + R)^N - 1]$$

Net profit
Net profit = gross margin - overhead

Odds of a series of events
Odds of a series of events = Odds of first event x odds of second event x odds of third event.....

Odds of a series, when each is the same
O = Odds   N = Number of events
Odds of a series when each is the same = O^N

Percentage increase/decrease
Percentage inc./dec. = (new figure - old figure) ÷ old figure

Percentage of a whole
Percentage of a whole = subgroup ÷ whole group

Percentile rank

Percentile rank = (Number of people at or below an individual score) ÷ (number of test takers)

Perimeter, square or rectangle

Perimeter = (2 x length) + (2 x width)

Price-earnings ratio

Price-earnings = market price/share ÷ earnings/share

Quick ratio

Quick ratio = cash ÷ liabilities

Return on assets

Return on assets = net income ÷ total assets

Return on equity

Return on equity = net income ÷ common equity

Speed, ending (if you know distance)

Ending speed = $\sqrt{2(\text{acceleration x distance})}$

Standard deviation

Subtract the mean from each score in the distribution. Square the resulting number for each score. Compute the mean for these numbers. This figure is called the variance.

Standard deviation = square root of the variance

Unemployment rate

Unemployment rate = (unemployed ÷ labor force) x 100

Velocity, ending

Ending velocity = (Acceleration x time) + starting velocity

Volume

Volume = length x width x height

z score

z score = (Raw score - mean) ÷ standard deviation

Answer Key

(Page numbers in parentheses indicate location of formula)

Chapter 2

1. **(p. 26)** Subtract the old figure from the new figure: 8,294 - 6,759 = 1,535
Then divide the difference by the old figure: 1,535 ÷ 6,759 = .227 = 22.7%

2. **(p. 26)** Use the same formula as above: 5,283 - 14,700 = -9,417
-9,417 ÷ 14,700 = -.64 = -64%

3. **(p. 27)** Divide the number of medal winners by the total participants: 227 ÷ 309 = .73 = 73%

4. **(p. 28)** 3% - 5% = -2% (It dropped 2 percentage points)
-2% ÷ 5% = -.4 = -40% (It dropped 40 percent)

5. **(p. 28)** 34.3 - 36.1 = -1.8 (It dropped 1.8 percentage points)
-1.8 ÷ 36.1 = -.05 = -5% (It dropped 5 percent)

6. The revenue from admissions and concessions dropped from 67 percent to 54 percent of the budget, so 67 - 54 = 13 percentage points. To find the percentage drop, divide -13 by the previous total, 67 percent: -13 ÷ 67 = -19%. Last year's county contribution was 46 percent, because the arts center earned 54 percent of of its budget from admissions and concessions. Forty-six percent of $1 million is $460,000. The year before the county contributed 33 percent of the $1 million, or $330,000. So last year the county paid: $460,000 - $330,000 = $130,000 more than the previous year. To find the percentage increase: $130,000 ÷ $330,000 = .39 = 39 percent

7. 4/5 = 4 ÷ 5 = 0.8 = 80 percent. So 80 percent of the streets were cleared.

8. **(p. 31)** This loan is compounded annually, not more frequently, so the bank will simply figure the interest when Amanda comes in to pay off the loan. Find the monthly interest by dividing the annual interest by 12: 6.5% ÷ 12 = .54% monthly interest. Now multiply by 10 to get the interest on 10 months: .54% x 10 = 5.4% Now figure her interest owed by changing 5.4 percent to a decimal and multiplying it by the loan amount: .054 x $6,300 = $340.20

9. **(p. 27)** Cindy's salary increase the first year of the contract: $24,000 x .025 = $600.
Her salary that year: $24,000 + $600 = $24,600. An alternative way to calculate this is to multiply the salary by 1.025: $24,000 x 1.025 = $24,600.
For subsequent years, start with the previous year's salary and plug in the new percentage. For year two: $24,600 x 1.035 = $25,461. Year three: $25,461 x 1.04 = $26,479.44.

10. **(p. 33)** Remember that you need to divide the rate by 12 before plugging it into the equation.
A = ($75,000 x [1 + .0028]^{60} x .0028) ÷ ([1 + .0028]^{60} - 1) = ($75,000 x 1.18 x .0028) ÷ (1.18 - 1) = 247.8 ÷ .18 = $1,375.67 is the monthly payment. There are 60 monthly payments, so the total amount the city will pay is $82,540.20. Subtract the original cost of the plow, $75,000, to find the total interest paid: $7,540.20

Chapter 3

1. **(p. 40)** Add up the numbers for each year: 7 + 92 + 53 + 81 + 46 + 19 + 34 + 51 + 6 + 25 = 414. Now divide by the total number of years: 414/10 = 41.4 is the mean.

2. **(p. 40)** To find the average (mean) pay you must take into account the number of workers employed in each trade. First, multiply the number of workers in a particular trade by the rate for that trade:

Detail workers: 18 X 7.98 = 143.64
Senior detail workers: 2 x 9.98 = 19.96

Large item workers: 6 x 5.48 = 32.88
Senior larger workers: 4 x 7.48 = 29.92
Level one supervisor: 2 x 11.49 = 22.98
Level two supervisor: 1 x 15.35 = 15.35

Now add all the amounts: 264.73. Divide that by the total number of workers (not just the different types of workers): 33. 264.73 ÷ 33 = $8.02 is the mean hourly pay of the workers.

3. **(p. 40)** Arrange the amounts in ascending order, then find the mid point. There is an even number of figures, so average the two middle figures: (12.28 + 16.75) ÷ 2 = 14.5 is median.

4. **(p. 43)** 116 x .60 = 69.6 This means 70 officers scored the same or lower than Jack Garfield. Since 116 officers took the test, 46 scored higher, so Jack didn't make the cut.

5. First subtract the mean, 71, from each school's score:
Rheingold: 56 - 71 = -15
Buchanan: 82 - 71 = 11
Bell: 76 - 71 = 5
Online: 69 - 71 = -2

Square each result:
Rheingold: 225
Buchanan: 121
Bell: 25
Online: 4
Calculate the mean for those numbers:
(225 + 121 + 25 + 4) ÷ 4 = 375 ÷ 4 = 93.75 This is the variance.
The standard deviation is the square root of the variance: 9.68

6. **(p. 45)** The larger the standard deviation, the more a group of figures varies from the mean. This indicates volatility. Thus, the Rocksolid Mutual Fund is more volatile than the Gold Platter Mutual Fund. If you were hoping to get rich quick, you'd likely want to take the risk of a fund whose value swings, hoping that you could get in on a down swing and sell on an up swing.

Answer Key

Answer Key 179

Thus, you'd buy Rocksolid. If you didn't want to risk your investment, you'd go with the fund that was more stable, Gold Platter.

7. **(p. 47)** Five people out of 540,000 are murdered per year, so divide by 5 and you get one out of 108,000. So the odds of any given resident of Fitztown getting murdered is one out of 108,000. All murders happened among adults, who are 75 percent of the population. The number of adults = 540,000 X .75 = 405,000. Only 1 out of 20 adults visits the red light district. One is 5 percent of 20, so 5 percent of the adult population visits the red light district: 405,000 x .05 = 20,250. All five murders happened among this small group of people. Dividing 20,250 by five converts the figure to one murder per 4,050 adults who vist the red light district.

Chapter 4

1. **(p. 60)** Monthly inflation rate = (131.6 - 130.4) ÷ 130.4 = .009 = 0.9%

2. **(p. 62)** To compare the two salaries you must convert the 1976 salary to a 2002 salary.
Target year salary = ($22,000 ÷ 57) x 179 = 69,087.72
So, accounting for inflation, the 1976 mayor earned more than the 2002 mayor.

3. **(p. 63)** Cost in one year = $95(1 + [.024 ÷ 12])^{12} = $95 x 1.024 = $97.31.

4. **(p. 65)** Trade balance = $30.8 billion exports - $12.7 billion imports = $18.1 billion surplus.

Chapter 5

1. **(p. 71** This is systematic random sampling.

2. **(p. 571)** Any sampling method could be used, but for this project the most likely would be cluster sampling or systematic random sampling. Both of these could work well with the relatively small campus population.

3a. **(p. 72)** 95% confidence level in a sample of 500 = a margin of error of 4.4%
99% confidence level in a sample of 500 = a margin of error of 5.7%
Were the results significant? Yes, because even taking the margins of error into account, the percentages don't overlap (excluding the "unknown" category).

3b. **(p. 72)** The margin of error was 3.1 percent.

3c. **(p. 72)** There is a small pattern in the middle with the differences getting smaller, then jumping because of the jump in population size. The smallest difference is 0.5 percent for 400 to 500. The largest is 4.4 for 100 to 200. The more numbers polled, the smaller the error.

4. a, b, c, d. n/a

Chapter 6

1. **(p. 87)** The gross margin on the sale is the selling price minus the original cost:
$13/ton - $10/ton = $3/ton. Tuckahoe has 330 extra tons that it sells to Gadsen, so its gross margin on that sale is 330 x $3 = $990

2. **(p. 87, 88)** First, figure the gross margin on PK Wholesale's sale to Tuckahoe: (1,000 tons x $10/ton) - (1,000 tons x $4.50/ton) = $10,000 - $4,500 = $5,500
PK's overhead during the three weeks in question is 3 x $1,256 = $3,768. Since the Tuckahoe sale was the company's only transaction during the three weeks, its net profit for that period is $5,500 - $3,768 = $1,732.

3a. **(p. 89)** NY Times total assets in 1999: $3,495,802,000 In 2000: $3,606,679,000
Total liabilities and stock holder's equity in 1999: $3,495,802,000 In 2000: $3,606,679,000. Why do they match? Because the stockholder's equity will always equal the value of assets minus liabilities, which balances the bottom line.

3b. No commercial paper outstanding in 1999, and no assets for sale in 2000.

3c. Stockholders' equity in 1999 = $1,448,658 ÷ $3,495,802 = .41 = 41 percent
Stockholders' equity in 2000 = $1,281,163 ÷ $3,606,679 = .36 = 36 percent
Conclusion? The stockholders were in better shape in 1999, because less of their company was committed to liabilities.

3d. In 1999 the company had a net total of $1,218,396,000 in property and $614,908,000 in current assets. In 2000 it had $1,207,160,000 in property and $610,766,000 in current assets. Both years it had more assets in property.

4a. Total revenue in 1999 was $3,156,756,000. Total revenue in 2000 was $3,489,455. The difference was $3,489,455,000 - $3,156,756,000 = $332,699,000. The percent difference was $332,699,000 ÷ $3,156,756,000 = .105 = 10.5 percent increase

4b. Percentage increase in revenues:
Newspapers: $3,160,247 - $2,857,380 = $302,867
$302,867 ÷ $2,857,380 = .106 = 10.6%
Broadcast: $160,297 - $150,130 = $10,167
$10,167 ÷ $150,130 = .068 = 6.8%
Magazines: $115,438 - $110,566 = $4,872
$4,872 ÷ $110,566 = .044 = 4.4%
New York Times Digital: $66,590 - $43,680 = $22,910
$22,910 ÷ $43,680 = .524 = 52.4%

So New York Times Digital showed the largest percentage increase in revenue.

4c. Net income in 1999 was $310,177,000. Net income in 2000 was $397,536,000. The percent increase was $397,536,000 - $310,177,000 = $87,359,000
$87,359,000 ÷ $310,177,000 = .282 = 28.2%

4d. Income taxes in 1999 = $228,287,000; In 2000 = $275,550,000

4e. New York Times Digital was not a profitable enterprise in 1999 or 2000, because it showed a loss each year. Its loss, as a percentage of sales, was 32 percent in 1999 and 105 percent in 2000.

Chapter 7

1. Determine the flucuation by finding the difference between the 52-week high and the 52-week low:
Aon = 44.80 - 13.30 = 31.50
Apache = 60.09 - 34.77 = 25.32
Aptinv = 51.46 - 40.06 = 11.40
Apagent = 26.52 - 16.87 = 9.65
Aon has fluctuated the most in the past year.
Aptinv has the highest price to earnings ratio.
Aon was the only stock that went up that day.
Apache went down the most that day.

2. Form 8-K

3. **(p. 105)** Current yield = (6% x 10,000) ÷ $9,500 = $600 ÷ $9,500 = .0632 = 6.32%

4. **(p. 105)** The interest rate for the bond remains the same, so the new current yield = (6% x 10,000) ÷ $9,000 = 6.7%

5. **(p. 106)** Bond cost (interest) = $1 million x 3.5% x 10 = $1 million x 0.035 x 10 = $350,000
The Village of Augusta will pay $350,000 in interest over the life of the bonds.

Chapter 8

1. **(p. 114)** Mill levy = $750,000 ÷ $25 million = 0.03 = 30 mills = $30 per $1,000 of assessed valuation

2. **(p. 115)** Assessed value = $210,000 x .75 = $157,500

3. **(p. 115)** Assessed value of the Bellmead house: ($325,000 x 0.9) - $20,000 = $272,500

Assessed value of the Andrews cabin: $50,000 x .25 = $12,500

4. **(p. 116)** Tax owed = $1.09 x ($56,000 ÷ 100) = $1.09 x $560 = $610.40

5. **(p. 116)** Tax in Millerton = $0.456 x ($100,000 ÷ $100) = $0.456 x $1,000 = $456
Tax in Smithville = $2.28 x ($100,000 ÷ $1,000) = $2.28 x $100 = $228

Chapter 9

1. **(p. 122)** One mile is 5,280 feet, and one nautical mile is 6,080 feet. So one nautical mile = 5,280 feet ÷ 6,080 feet = .87 miles
So the distance from shore to ship is 112 miles x .87 miles/nautical mile = 97.4 nautical miles
The supply boat travels at 15 knots, which is 15 nautical miles per hour. Turn the above conversion around: one nautical mile = 6,080 feet ÷ 5,280 feet = 1.15 miles.
Thus, 15 nautical miles/hour x 1.15 nautical miles/mile = 17.25 miles/hour.

2. **(p. 123)** First, convert 4 hours and 34 minutes to a decimal by dividing the total number of minutes ([4 hours x 60 minutes/hour] + 34 minutes = 274 minutes) by 60: 274 ÷ 60 = 4.57 hours. Now use the rate formula: rate (speed) = 26.2 miles ÷ 4.57 hours = 5.73 miles/hour

3. **(p. 128)** Momentum = 785 kilograms x 48 kilometers/hour = 37,680 kg/k/hr

4. **(p. 128)** The key to this problem is keeping the units of measure consistent. The acceleration, 1.5 miles/hr/sec, needs to be changed to miles/hr/hr. There are 3,600 seconds in an hour, so the acceleration = 3,600 sec/hr x 1.5 miles/hr/sec = 5,400 miles/hr/hr. Now use the formula for ending speed on page 102, which works when one knows the acceleration and distance, but not the time.

Ending speed = $\sqrt{2(5,400 \text{ miles/hr/hr} \times 0.1 \text{ miles})}$ = $\sqrt{1080}$ = 32.9. The car was moving about 33 miles per hour when it hit the wall.

5. **(p. 125)** Acceleration = (123 mph - 81 mph) ÷ 2 minutes = 42 mph ÷ 2 minutes = 21 mph/minute.

6. **(p. 126)** The camera is moving downwards due to gravity. The acceleration of gravity is 9.8 meters/sec/sec.

Ending speed = $\sqrt{2(9.8 \text{ m/s/s} \times 50 \text{ m})}$ = $\sqrt{980}$ = 31.3 m/s
The camera's momentum upon hitting the ground was:
Momentum = 20 kilograms x 31.3 m/s = 626 kg/m/s

Chapter 10

1. **(p. 135)** Perimeter = (2 x 120 yards) + (2 x 50 yards) = 240 yards + 100 yards = 340 yards
Area = 120 yards x 50 yards = 6,000 square yards

2. **(p. 135)** A square has four equal sides, so if the perimeter is 56 inches, each side must be 56 ÷ 4 = 14 inches.

3. **(p. 135)** The new forest preserve is a rectangle. Two sides are 11 miles long, and the other two sides are 31 miles long. Thus, the length of the total perimeter would be (2 x 11) + (2 x 31) = 22 + 62 = 84. But the road in question will only run along the three sides of the preserve not bounded by the river, so the

length of the road will be (2 x 11) + 31 = 22 + 31 = 53 miles.

4. **(p. 137)** The area of the tract of land is: Area = 6 miles x 1.2 miles = 7.2 square miles. One square mile equals 640 acres, so the tract = 7.2 square miles x 640 acres/square mile = 4,608 acres. There are 43,560 square feet in an acre, so the tract = 4,608 acres x 43,560 square feet/acre = 200.7 million square feet.

5. **(p. 136)** Area of a triangle = .5(23 ft x 42 ft) = .5 x 966 square feet = 483 square feet

6. **(p. 135, 137)** The area of the lot = .5(500 ft x 1,250 ft) = 312,500 square feet. Convert this to acres: 312,500 square feet ÷ 43,560 square feet/acre = 7.17 acres. Since each acre costs $20,000, the lot is worth 7.17 x $20,000 = $143,400

Chapter 11

1. **(p. 142)** 302 children x 8 oz x 5 days = 12,080 oz per week. Convert to gallons: 12,080 oz per week ÷ 128 oz per gallon = 94.38 gallons per week.

2. **(p. 142)** Assume the source in the United States is measuring its oil in U.S. gallons, and the sources in Canada and Saudi Arabia are measuring oil in imperial gallons. One U.S. gallon is only 4/5 of an imperial gallon, which means each imperial gallon is 5/4 of a U.S. gallon. In order to compare prices, change the U.S. source measurement to imperial gallons. First change 5/4 to a decimal: 5 ÷ 4 = 1.25. If each U.S. gallon costs 59 cents, then each imperial gallon would cost 59 cents/gallon x 1.25 gallons/imperial gallon = 73.75 cents per imperial gallon. Now we can compare prices, and see that the oil in Canada, at 39 cents per imperial gallon, is the least expensive.

3. **(p. 144)** Volume = 6.5 feet x 10 feet x 8 feet = 520 cubic feet

4. **(p. 144)** 22,000 pounds ÷ 2,000 pounds/short ton = 11 short tons
11 short tons x .89 long tons/short ton = 9.79 long tons

Chapter 12

1. **(p. 150)** First convert 5' 5" to a decimal. There are 12 inches in a foot, so 5 inches ÷ 12 inches/foot = .42 feet. So Sara is 5.42 feet tall. Convert that to meters: 5.42 feet x 0.3 meters/foot = 1.63 meters.

2. **(p. 150)** The debris scattered: 2.6 meters x 3.3 feet/meter = 8.58 feet.

3. **(p. 154)** Gill's research indicated the 50-gallon tank was filled twice a week, equaling 100 gallons used per week. The alderman reported that the tank was 150 liters, and was being filled 3 times a week, for a total of 450 liters. In order to compare, convert the units in Gill's measurement to liters: 100 gallons x 3.8 liters/gallon = 380 liters. That is a difference of 70 liters.

4. **(p. 152)** 4,000 square yards x 0.8 square meters/square yards = 3,200 square meters

5. **(p. 153)** 15 short tons x 0.9 metric tons/short tons = 13.5 metric tons

6. **(p. 154)** 2.25 kilograms x 2.2 pounds/kilogram = 4.95 pounds, or almost 5 pounds

7. **(p. 155)** 58 degrees Celsius = (1.8 x 58 degrees Celsius) + 32 = 136.4 degrees Fahrenheit.

Appendix 1

1. **(p. 162)** Yes, the both equal 15

2. **(p. 162)** Do the equation inside the parentheses first: 4 + 13(7 + 5) = 4 + 13(12) = 4 + 156 = 160

3. **(p. 163)** The square root of 25 is 5, so Purnell will donate $5 million this year. The square root of 36 is 6, so Purnell will donate $6 million to the museum when his daughter is 36.

4. **(p.163)** 6 to the third power is the same as 6 x 6 x 6 = 216

5. **(p. 164)** Since the denominators are equal, it's easy to add the fractions. 1/5 + 2/5 = 3/5, so the siblings own 3/5 of the grocery story, leaving 2/5 for the other partners.

6. **(p. 164)** First calculate how much Shameeka owns after she buys Colin's share. You must make the denominators equal, so multiply them together. Then calculate the new numerators by multiplying the denominator of each fraction by the numerator of the other fraction.

Shameeka owns 1/5 + 1/2 = 2/10 + 5/10 = 7/10, or 70 percent.

This leaves 3/10 for the other partners, or 30 percent.